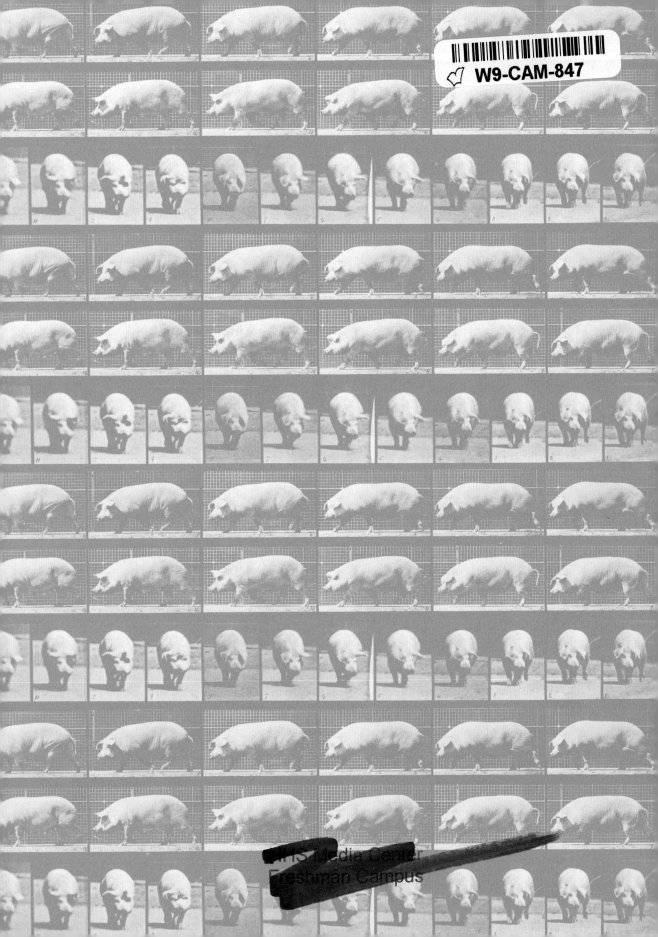

THE WHOLE HOG

LYALL WATSON

THE

WHOLE HOG

EXPLORING the EXTRAORDINARY
POTENTIAL of PIGS

SMITHSONIAN BOOKS
WASHINGTON

© 2004 by Lyall Watson
Photographs copyright as listed on page 206
All rights reserved
Published 2004 in the United Kingdom by Profile Books

Copy editor: Julie Carlson
Production editor: Joanne Reams
Designer: Empire Design Studio

Library of Congress Cataloging-in-Publication Data
Watson, Lyall
      The whole hog: exploring the extraordinary potential of pigs / Lyall Watson.
            p. cm.
      Includes bibliographical references (p.) and index.
      ISBN 1-58834-216-6
            1. Suidae. 2. Suidae—Behavior. 3. Human-animal relationships. I. Title

      QL737.U58W28 2004
      599.63'3—dc22   2004052248

British Library Cataloging-in-Publication Data available

Manufactured in China not at government expense
10 09 08 07 06 05 04   1 2 3 4 5

For Alice, with love

Contents

The Cast

AMERICAN PIGS

Collared Peccary
Tayassu tajacu

White-lipped Peccary
Tayassu pecari

Giant Peccary
Catagonus wagneri

DOMESTIC PIGS

All breeds
Sus scrofa domesticus

EURASIAN PIGS

Wild Boar
Sus scrofa

Pigmy Hog
Sus salvanius

ISLAND PIGS

Philippine Warty Pig
Sus philippensis

Visayan Warty Pig
Sus cebifrons

Sulawesi Warty Pig
Sus celebensis

Javan Warty Pig
Sus verrucosus

Bearded Pig
Sus barbatus

Babirusa
Babyrousa babyrussa

AFRICAN PIGS

Bush Pig
Potamochoerus larvatus

Red River Hog
Potamochoerus porcus

Forest Hog
Hylochoerus meinertzhageni

Desert Warthog
Phacochoerus aethiopicus

Common Warthog
Phacochoerus africanus

In a Pig's Eye

I have had close relationships with three species of wild pigs, each a chance encounter on a different continent, and all continue to enrich my life in surprising ways. I know of no other animals that are more consistently curious, more willing to explore new experiences, more ready to meet the world with open-mouthed enthusiasm. Pigs, I have discovered, are incurable optimists and get a big kick out of just being. We have a lot to learn from them, and it is in this spirit that I offer everything that follows.

EURASIAN BOAR HAVE THE largest range of any wild ungulates and are the direct ancestors of most domestic pigs.

BERKSHIRE PIG ($\frac{1}{16}$ nat. size).

PIGS MAY HAVE DOMESTICATED themselves long before we settled down to an agricultural way of life. This is not a textbook or even a comprehensive field guide. It is a personal inquiry, an attempt to bring together everything about wild and domestic pigs that might counteract some of our misguided perceptions about them and shed some light instead on creatures a lot like ourselves, whose most useful tools seem to be their brains.

The First Leg of my exploration begins with defining "pigness" and investigating pig history and natural history with the help of three species that lived near my childhood home in Africa—the bushpig, the forest hog, and the inimitable warthog. These three pigs present a vivid demonstration of the steps that pigs have had to take from grazing to browsing to the creative benefits of being completely omnivorous.

The Second Leg introduces three New World wild pigs—the collared peccary, the white-lipped peccary, and the giant peccary—and uses their refined senses of smell, sound, and community to show how swine society has arisen and found its place in a wide variety of ecological niches.

The Third Leg looks at the history of human/pig relationships, examining evidence to suggest that pigs, in the form of the Eurasian wild boar, may have domesticated themselves long before we settled down to an agricultural way of life in company with our other "best friend," the dog. Pigs brought a far more generous package of benefits to their merger, including the fact that they taste a lot better than dogs do, and this association can be traced from its beginning almost 30,000 years ago through Asia and Europe to the Hog Belt of the American Midwest.

The Fourth and last Leg tackles the mystery of pig ascendancy and intelligence, showing how profoundly pigs have captured our imagination, starting with ancient Celtic boar cults, going on through Gadarene swine and ecclesiastical sows to pigs both sacred and profane in literature and religion. No other group of living things has been so often discussed and so little understood, and disagreement about this is implacable. The taboo against pigs in the Middle East is just as vehemently opposed by the reverence in which pigs are held in most of the Pacific, where they have become symbols of political and social power. Anyone there who refuses to eat pork is regarded as inhuman.

This final chapter introduces the extraordinary babirusa—the "deer pig" of Sulawesi—and all five species of warty pigs that survive in the Malay Archipelago. Remnants of their ancestors have been found together with human remains 40,000 years old, endorsing the depth of our mutual association and highlighting the existence of what has come to be called "pig culture"—a tendency for pigs to form amiable, accommodating societies among themselves, and for humans to find their company not just rewarding, but of enormous ritual and social significance.

And the Conclusion assesses what science has begun to discover about pig intelligence, and why this matters.

It matters a great deal to me, but my enthusiasm for pigs has surprised most of those who know me and has forced me to justify my decision to write a whole book about them. It goes something like this.

All animals are not equal, not even for a biologist like myself. Some, whether we like it or not, are more equal, more interesting, more likely to grasp the imagination. It is not always easy to identify what it is that gives them this advantage, this happy accident that catches our interest, for natural selection works as randomly and heedlessly on the world of ideas as it does on amino acids.

For a start, pigs have it, and sheep don't.

It was no accident that George Orwell cast pigs as the rulers of his farmyard. The task of organizing the others "fell naturally upon the pigs, who were generally recognized as being the cleverest of the animals," while the sheep were content to lie about in the fields bleating "Four legs good, two legs bad! Four legs good, two legs bad!" for hours on end.

At one level, *Animal Farm* is a powerful political tract and a scathing satire on human folly, but it is also, and just as profoundly, a very effective fable in the Aesop tradition. In this allegorical lesson, the feuding "big pigs" Napoleon and Snowball are unmistakably based on Stalin and Trotsky as part of an attack on Soviet Communism at the time of the Russian Revolution, but beneath the zoomorphism of the story that served his ideological purpose, Orwell displays an acute awareness of biological design.

His choice, from all the hundreds of domestic pig breeds, to make Napoleon "a large, rather fierce-looking Berkshire boar" is inspired. This breed is indeed hardy, rugged, and muscular, predominantly black-colored with large upright ears, a slightly dished face, and a reputation for being somewhat stubborn. Since its breed book was opened in the mid-nineteenth century, the Berkshire (and its crosses) has spread from England to Europe, most of Asia, and all of North America, where it has become known as "everyone's pig." Just the sort of hog you would expect to get up on its own hindlegs.

No one who has read *Animal Farm* can ever again sit through a diatribe that pays lip service to "equality" without hearing an echo of the essential caveat, "but some are more equal than others." And close behind this admonition comes Orwell's supplementary caution—"but the pigs were so clever that they could think of a way round every difficulty." True, and as useful in pig husbandry as it is in politics.

While researching this book, and still struggling to explain my curiosity about pigs and their alleged intelligence, I visited a pig farm in England on which there were several rare breeds. They were kept in relatively pleasant circumstances that allowed them to spread themselves out and be elective, rather than obligatory, wallowers, choosing to get mucky or not. Under these conditions, it became obvious that pigs can be surprisingly fastidious.

This was brought home to me by a solitary Tamworth boar, an aristocrat of his kind, with a pedigree that goes directly back to the Eurasian wild boar. He and his breed are lean, rangy pigs with long heads and impressive snouts, born foragers in bright, ginger-colored coats. And this one was a fashion-plate pig, washed and oiled and curry-combed, sitting back on his haunches like a lord.

He caught my eye and held it in a way very few animals can or ever do. Winston Churchill, in one of his many rural musings, said: "Cats look down on you; dogs look up to you; but pigs look you in the eye as equals." This one did, making scrutiny mutual, and that stopped me in my tracks.

The facial muscles of suids concentrate largely on opening and closing the mouth, often with frightening strength and speed, but they cannot compete with the rowdy variety of expression that became available to primates when our faces flattened and our whiskers shortened, and the old muscles were freed for more expressive purposes. Pigs, on the whole, are poker-faced, giving little away, but they can be surprisingly eloquent, particularly around the eyes, and in this pig's eye I saw: "Well, what did you expect? More swinery? Nothing but filth, sloth, and gluttony?"

It is more than a little disconcerting to be frowned at by a pig, and it reminded me of another occasion on which I found myself on the receiving end of a sharp non-human gaze. It happened at the London Zoo where I once spent three glorious years studying for a doctorate in ethology with Desmond Morris.

During lunch breaks, I made a habit of strolling through the collection, enjoying the huge variety on display along a route that somehow always ended outside the Ape House where its oldest resident, Guy the gorilla, held open-air court.

By that time, in the early 1960s, he had been at the zoo for sixteen years, most of them alone because he had become "mispithic," given to beating up every partner the keepers provided. He seemed resigned to a monastic existence and spent a lot of his time between meals just sitting in the sun thinking deep anthropoid thoughts.

On that day, he was lying comfortably on his back, eyes open, head pillowed on one huge upper arm, at peace with his world. Then he allowed himself a small, sleepy yawn that, in the manner of yawns, was followed immediately by a far larger and more satisfying one that brought him up to a sitting position with both arms outstretched. This great yawn exposed every yellow tooth in his cavernous

mouth and ended with the back of one hand covering his gape in a gesture so natural that, for a full minute, I didn't find it incongruous.

MOST PIGS ARE POKER FACED, giving away little, but they can be surprisingly eloquent around the eyes and snout.

Then the significance of the act caught up with me. Covering your mouth when you yawn isn't an innate gorilla gesture. It is a cultural nicety, one of those strange social things that we have had to learn and that is, or so I thought, uniquely human.

I gawped at Guy until something about my intensity caught his attention and he took a long hard look at me, as if to say: "Well, what did you expect? I *am* a great ape!"

He was. And I now feel doubly honored, having been at the receiving end of an equally knowing and censorious glare from a gorilla and a boar, both of whom not only looked at me but left me with a feeling that I was being *watched*. Guy's regard was not totally unexpected—he was a close relative after all—but the Tamworth boar's stare was unnerving.

I have since learned that pigs are past masters of the art. They grow up on the game of "Who Blinks First?" and can hold their ground against anyone. Each time I join a herd of pigs, in captivity or in the wild, the same thing happens. Everything stops. All foraging, grunting, and blowing, every boisterous game, comes grinding to a halt and suddenly I am the center of silent attention, the focus of every eye. Ears may twitch a little, but no one bats an eyelid or breaks visual contact until I concede defeat, saying something self-conscious like: "What?"

The wise old Tamworth had my number right from the start. He knew I was an easy mark. The moment I remembered Guy's gaze and made that connection, I relaxed and let my breath out in a soft, open-mouthed pant, which, as it happens, was precisely the right thing to do.

The boar's response was instantaneous and reciprocal. He panted back, and I remembered that this was the way a pet bushpig in Africa had greeted me as a child. I replied once more in kind, and when the boar's long look softened, I knew for certain what I had already begun to suspect. I was looking at an ape in swine's clothing.

PLATE XLVII.

BOAR. TROTTING.

THE STRUCTURE AND DESIGN OF pigs have changed very little during the last 40 million years, but something interesting has happened inside their heads.

Pigs may be Artiodactyls—even-toed, hoofed animals closely related to sheep, goats, deer, and cattle—but they have no equals in the order.

Pigs are different. They think, work, and play outside the pen. Their structure and basic design have changed very little during the last 40 million years, but something interesting has happened inside their heads, something that makes them the most engagingly idiosyncratic of all ungulates.

The sad truth, however, is that we know very little about pigs; and little we think we know about pigs is true.

Biologists acknowledge that there are many things about pigs we just have to guess at, because we see so little of them in the wild, witnessing no more than a tiny fraction of their lives. Even in captivity, pigs are seldom given the chance to show us what they can truly do. Most of the research on domestic pigs is market-driven, governed largely by concerns about productivity. And the analysis of what pigs are is often untidy, uninspired, and inclined more to what we want, or imagine, pigs to be.

So pigs remain mysterious. Few animals are the object of more deeply rooted misapprehensions, or more passionate justifications. And none carry a greater burden of semantic confusion, being simultaneously described as "beautiful" and "hideous," "greedy" and "restrained," "clever" and "ignorant," "pure of heart" and "instruments of the devil."

Pigs make us uneasy. They return our scrutiny and never give an inch. They know what's happening and let us know what's happening. They are out of our control, and that leaves us with uncomfortable feelings that find expression in more superstitions, ceremonies, folktales, and taboos than we attach to any other creature.

The way in which pigs have become repositories for all our own most foul and bestial attributes is, however, a dead giveaway. Pigs, more than anything else, remind us of ourselves—and that is why I find them so fascinating.

So, here goes . . .

The creatures outside looked from pig to man, and from man to pig and from pig to man again; but already it was impossible to say which was which.

—GEORGE ORWELL, *Animal Farm*, 1945

THE FIRST LEG

The Whole Hog

Pigs grunt in a wet wallow-bath, and smile as they snort and dream. They dream of the acorned swill of the world, the rooting for pig-fruit, the bagpipe dugs of the mother sow, the squeal and snuffle of yesses of the women pigs in rut. They mud-bask and snout in the pig-loving sun; their tails curl; they rollick and slobber and snore to deep, snug, after-swill sleep.

AN AURA OF MYSTERY SURROUNDS pigs. They are slippery creatures, literally hard to pin down but undeniably smart.

—DYLAN THOMAS, *Under Milk Wood*, 1954

There are words in every human language that seem to have been crafted with pigs in mind, words like "grunt" and "snuffle," "snort" and "slobber" in English. And when these have served their purpose, poets such as Dylan Thomas have no trouble in finding others that can be hyphenated to fit the design, perhaps because the qualities they depict are as familiar to us as they are apt for swine.

In truth, there are so many easy words available for describing pigs, and what they do, that it becomes necessary to discard some, and define others out of useful existence.

The *Oxford English Dictionary* decides that "pig" itself is etymologically obscure but probably derives from Old English *pigg, pigge or pygge* and was originally applied to the "young of a swine." "Swine," in its turn, is more Teutonic, from *suine, swiyne,* or *swynne,* and was most properly used for domestic animals.

Low German and Early Modern Dutch derivatives such as *pigga* and *vigghe* merely add to the confusion. As do claims that "pig" today can be applied as easily to: an earthenware vessel, a sixpenny bit, a block of salt, a parcel of hemp, a printer, an iron ingot or a piece of lead for ballast, something sent through a pipeline, a small cushion, a male chauvinist, and a large policeman, usually one in plain clothes.

A pig is a pig for all that, but for my purposes I choose "pig" to mean any member of any age, in any country or condition of the lineage of Suiformes—as long as it is not a hippopotamus. That means any even-toed ungulate that isn't aquatic, doesn't ruminate, and does have old-fashioned, low-crowned teeth. That's about as simple as I can make it, but pigs always start out looking simple and tend to get complex rather quickly.

In Latin, things are a little easier. A pig is *porcus,* which not only defines the animal but also describes its meat and at least one of its species. And the stem genus in the whole group is *Sus,* from the Latin for a sow. This usage led to Anglo-Saxon *sugu* for a sow, Cornish *hoch* for a boar, and Irish *suig* for both sexes—a form that was carried across the Atlantic in the sixteenth century as "hog," which is still how most Americans describe any animal that resembles the lean, long-snouted, razor-backed descendants of the Eurasian wild boar that were introduced to the New World by the first Spanish colonists.

Putting aside the fact that "hogs" now also apply to ten-cent coins in the United States, motorcycles, and anyone who behaves in a greedy fashion, I can only admit that the confusion between "pig" and "hog" remains, and is enshrined in, the common names for wild suids. The largest and the smallest ones are hogs—forest hogs and pigmy hogs—while the rest are pigs. And the two very closely related species in the genus *Potamochoerus* are described variously as bushpigs and red river hogs.

There is no logic in it at all, and the best I can do is to point out that, as far as I am concerned, all suids are pigs unless they are forest, pigmy, or red river hogs; or belong to any one of three species confined to the New World, where they have always been known as *pecari* or peccaries, meaning "animals which make many paths through the forest."

And while we are on the subject, it may help to know that within any wild species or domestic breed, *boars* are adult males; *sows* are breeding females; *barrows* are castrated male eunuchs; *gilts* are virgin females; *shoats* are adolescent, newly weaned males; and the rest, infants of both sexes up to the age of weaning, are *suckling pigs* or *piglets.*

IF I HAD TO CHOOSE JUST ONE WORD TO SUM UP THE NATURE OF PIGS, IT WOULD HAVE TO BE "GREGARIOUS." Pigs are highly social, living in family groups that maintain close contact and a gamy kind of togetherness that we associate more with primates than with ungulates. They are intensely aware of each other at all times, keeping in touch with a concert of small agreeable sounds, the sort of sounds that always get answered and help maintain group structure, even when they are out of sight of one another in dense undergrowth. It is not for nothing that such tight little societies are called "sounders."

At the heart of each sounder is the "mother family," a basic social unit consisting of a sow and her litter who stay attached until, and sometimes even after, the next litter is born. Often they band together with other sows and their young, forming a harem for a single boar who must fight late each autumn to defend this extended family.

Young females stay with their mothers until they become sexually mature in their second year and have litters of their own. Young males leave the group voluntarily, or are chased out when they come of age, and are unlikely to mate until their fourth year. In the meantime, they join loose, satellite groups of sub-adults, waiting their turn to challenge a harem-keeper.

In some species, boars remain in close contact with the sounder, but old males tend to drift off on their own. Old sows never live alone but linger on in the society, becoming matriarchal figures whose presence often determines where a group chooses to forage, and when they should move on. In all of which, they bear a marked resemblance to elephant society.

Pigs are seldom territorial, but each sounder operates within a shifting home range that may overlap with other related sounders in wider associations or "clans." And every home range includes a number of fixed points connected to each other by a network of paths.

The first of these fixed points is a resting place. Pigs are legendary and awesome sleepers, spending roughly half of every twenty-four hours slumped in comfortable immobility. Half of this is deep, slow-wave sleep with occasional snoring, interrupted by short periods of paradoxical, rapid-eye-movement sleep that fosters dreaming. Piglets seem to do more of this than older animals. The other half is spent in an unusual, half-awake, drowsy state that probably includes enough awareness to be semi-alert. But before pigs can enjoy any such idleness, and very clearly they do enjoy it, they need to find or make a suitably secure shelter.

Some pigs build nests, which vary from temporary grassy hollows to well-constructed hides in which each sheaf of leaves is meticulously placed to provide both comfort and security. Such ambitious structures are usually for farrowing, but even an afternoon nap requires some semblance of order and involves stereotypical behavior, perhaps a little ritual rooting and turning, before it is possible to settle down.

Lying down itself is a suid ceremony. Most pigs begin by sitting doglike on their hams, then they lower their forelegs, and finally they collapse, falling on their sides with "whumphing" sounds of satisfaction. Others kneel first on their forelimbs, holding that position briefly, as if genuflecting, before settling their heavy rears so that they end up with legs gathered beneath them and chins on the ground.

Some species rest and sleep together in amiable clumps of bodies in close contact with one another, partly because their spines are too rigid to allow them to curl up like carnivores, but mainly because they are naturally contact animals and need to keep warm in cold weather. Pigs huddle in ways that are seldom seen among other ungulates.

Our experience of wild pigs suggests that they are nocturnal, but wherever they exist without hunting pressures, they seem to relax into a crepuscular way of life. They forage and travel and interact most often in the early morning and late evening, when it is not too hot or too cold and when humid air carries its greatest load of environmental odors. If anything, the natural pattern of pig activity is a diurnal one, confirmed by the fact that they have retinas with color-sensitive cone cells almost as good as our own.

The second fixed point in a pig's life is a scratching post of some kind. Because of their short, muscled necks, pigs find it very difficult to groom themselves. They can reach some parts with their hind feet, but for the rest they rely on being able to find a convenient tree stump, rock, or anthill against which they can rub themselves after wallowing. Failing that, most pigs rely on the generosity of other pigs in rub-downs and combings with snouts and incisor teeth.

These attentions depend on goodwill and reciprocity that cannot be guaranteed, but peccaries have taken these understandings a crucial step further by building mutual grooming into their most common greeting ceremony. They stand, facing in opposite directions with sides touching, while each animal rubs the side of its head vigorously against the other's hindquarters. This has become a peccarine way of shaking hands and may have begun for the same purpose—an exchange of olfactory signals at close quarters.

The third essential landmark in any pig's home range is a place for defecation, a "dunging site" where all members of a sounder can, and usually do, deposit their feces. Such dung heaps are very common among ungulates but remain poorly understood. They appear to be hygienic and are always firmly set apart from designated eating and sleeping areas, even in domestic circumstances. In the wild, they may also serve as territorial markers or olfactory notice-boards. They probably have a number of functions, but on a casual survey of the haunts of several pig species, I have been touched to discover that such deposits are very often discreetly tucked, out of sight and personal shame, under shrubs or tangles of vegetation. In short, they are pig privies.

This doesn't surprise me. One of the first things I ever learned about pigs was how extraordinarily fussy they are about being toilet-trained.

AN AURA OF MYSTERY SURROUNDS PIGS. THEY ARE SLIPPERY CREATURES, LITERALLY AND metaphorically hard to pin down. With the exception of their splendid snouts, they are structurally primitive, unspecialized and unrepentantly primal. And yet by general agreement, they are regarded as highly intelligent, up there with apes and dolphins. Few other animals can match them for their combination of simplicity and complexity.

I find such contradictions alluring. They demand investigation, and what I hope to be able to provide here is a fresh look at an animal that we haven't really investigated, not because it is too rare or too difficult or too dangerous, but simply because it has been too close for us to see with any clarity.

Pigs have been subject to the same selective pressures as all the other ungulates, but at every fork in the road they seem to have made the sort of canny decisions that would give them the best chance of adapting to changing circumstances.

Proto-pigs, going back about 40 million years, were scary. They include *Entelodont,* a scavenger with a head over three feet long, decorated with knobby flanges wrapped around a tiny brain. These rangy animals were largely carrion feeders, but judging from their large, well-worn teeth, they were probably more or less omnivorous. It is interesting to know that ancestral pigs were so versatile. This makes sense of the very catholic tastes of pigs today.

The skulls of entelodonts bear the unmistakable signs of severe head wounds inflicted upon one another, but the lack of crushed skulls among the fossils, and the fact that eyes and noses were seldom damaged, suggests that aggression was inhibited, as it still is among living pigs, by the development of stylized patterns of dominance and submission.

Entelodonts first appeared in Asia but soon migrated to Europe and the Americas, taking up all the territories still occupied by modern wild pigs. One fossil recently discovered in Nebraska has been called *Dinohyus*—"the Terrible Pig"—because it was buffalo-sized with a crest of shaggy erectile hair running down its neck, slabs of muscle across its shoulders, and a set of long, sharp teeth. This monster probably lived at the carnivorous extreme of ancestral pigs, because by 30 million years ago, most species were smaller and much more agile.

They walked, not on clumpy hoofs like the entelodonts, but on the points of cloven feet. But even during the transition necessary for greater speed, pigs never committed themselves entirely to the unguligrade condition. While finding their teeth, they hedged their bets and retained evidence of their history in the form of "false hoofs" or "dew claws" that hung behind their feet as "trotters," leaving tracks in soft soil and keeping their options open.

They still do. This conservative approach, with

BUSHPIGS ARE THE MOST COLORFUL pigs in the world, demonstrating their genetic diversity by producing four distinct color forms.

versatile teeth and toes, has ensured that pigs have never strayed too far into any evolutionary opportunity. They may have hoofs, but their legs are still supported by twenty-five of the original thirty-one bones. Once the need for speed was satisfied, giving pigs a fair chance of getting away from predators at speeds of up to 30 miles an hour, they kept enough of their relatively short, unspecialized limbs to remain capable of simple manipulation. Pigs of all kinds routinely use one of their fore feet to hold something still while they examine, shred, or eat it with great skill, something deer and cattle and camels seem unable to accomplish.

Heaven help us now if hogs had kept all five fingers.

There are three different kinds of wild pig in Africa.

Bushpigs, of the genus *Potamochoerus*, comprise two living species, both rain forest forms, that are superficially similar to the core genus Sus, which covers all of Eurasia. They are generalized omnivores. Then there is a single species of forest hog belonging to the genus *Hylochoerus*, a large, archaic, gallery forest pig that has partially specialized as a browser. And finally there are two warthogs of the genus *Phacochoerus*. These are highly adapted, arid savanna pigs that have specialized as grazers.

All three genera are survivors of a lineage that has spread throughout the continental tropics during the last 40 million years. They have few immediate ancestors in common and, apart from their obviously different external characters, can all be separated easily by the nature of their teeth.

Bushpigs of the genus *Potamochoerus*— from the Greek *potamos* and *khoiros* for "river pig"—are the smallest and least specialized of the three African pig genera.

They are widespread in rainforest and all the moister parts of tropical Africa that provide river courses, reed beds, swamps, dense undergrowth, and tall grass on rich soft soil.

Bushpigs are compact, short-legged, and shaggy, with little indication of a neck. Their long, narrow heads represent almost a third of their total length and sport leaf-shaped ears and an erectile mane of coarse hair that runs all the way down the spine. Their tails are long and hairless, except for a tuft on the end.

They are the most colorful pigs in the world and come in two species. *Potamochoerus larvatus*—from the Greek for "having a mask"—has light gray facial hair and a coat that varies from honey blond to reddish-gray and black. *Potamochoerus porcus*—from the Latin for "pig"—has a narrow white crest on its back, white eyebrows, and long white tufts of stiff hair on its cheeks and the tips of its ears. But its most startling feature in some areas is a smooth coat of beautiful russet color, which gives it the common name of red river hog. *P. porcus* is less common than *P. larvatus,* being confined to and persecuted in the vanishing rainforests of Gambia and Zaire in Central Africa.

Both species feed on bulbs, roots, worms, insects, and occasional carrion and follow chimpanzees, baboons, and birds to fruits in season. And both are known, in hard times, to drive leopards from their kills. Red river hogs have also been seen moving large boulders in creeks to catch crabs.

In South and East Africa, bushpigs leave the fringes of the forest in the cool dry winter months and subsist largely on the roots of woody plants, but the favored food of both species is fruit. They upset farmers by uprooting banana and papaya trees to get at ripe fruit, and they create havoc in crops of corn and melons; but on the credit side, their rooting and furrowing activities in search of grubs control insect pests, break up the soil, encourage natural mulching, and distribute seeds in segmented, cylindrical droppings.

Bushpigs rest during the heat of the day in self-excavated nests of the densest undergrowth, and they give birth in such retreats to an average of four piglets after 120 days of gestation. The young are generously decorated with broken stripes that make them almost impossible to see when they lie absolutely still, "playing possum" on the dappled forest floor at the first sign of danger. They are weaned at six months and follow the mother until she actively chases them out into sounder society.

There are no marked differences in size between the adult sexes, both growing to an average length of 4 feet and an average weight of about 160 pounds, but big boars have bony ridges on their snouts, and lower canine teeth that grow into 3-inch tusks. These come into play in fights with other males who strut around each other, manes bristling as they champ their jaws and paw the ground, releasing aggressive phero-monal odors from their salivary glands. Battles during courtship and tussles for dominance are carefully inhibited by submissive displays in which ears stand up straight in unmistakable fright. But in encounters with alien males or outside threats, bushpigs of both sexes are fearless, deliv-ering lethal gashes with their razor-sharp tusks or tushes. All bushpigs also carry the full complement of forty-two pig teeth.

Sounders vary in size from about a dozen up to temporary associations of more than fifty individuals. These latter may range over several miles each day as the enlarged group exploits the possibilities of a loose terri-tory of anything up to 10 square miles. Red river hogs, in particular, will cross rivers without hesitation, but in fact both species are strong swimmers. When the Kariba Dam on the Zambezi River first filled, many bushpigs appeared to be marooned on isolated islands. They were caught by "rescuers" of Operation Noah and released on dry land several miles away, only for the rescuers to discover that the pigs were back on the same islands the following day. Some of them were even seen hiding underwater with just their noses resting on a convenient branch.

Both sexes of bushpig have glands on their feet, on the nape of their necks, and in the corners of their eyes, and they use them to leave scent marks on each other and on rubbing posts. Males of both species also have huge tusk glands in pouches at the base of the upper canine tooth, and I have seen boars using their tusks to put fragrant gashes on trees as high as they can

reach, presumably as territorial advertisements. Most sounders proclaim their presence in a habitat by letting their droppings accumulate in established "latrines."

Little is known about bushpig communication beyond the fact that they have a wide range of grunts and squeals that signal contact, alarm, assembly, and flight. Though no specific studies have been made on their cog-nition, anyone who has had the chance to observe bush-pigs in the wild comes away with a feeling of respect for them, so much so that these unspecialized suids have been described as "sagacious and intelligent."

IN SOME WAYS, BUSHPIGS ARE THE MOST PRIMITIVE, unspecialized members of their family, with single stomachs and legs made up of twenty-five bones. Fossil discoveries suggest that *Potamochoerus* may be a direct descendant of the stem form *Palaeochoerus,* the "Old Pig" of the Oligocene, 30 million years ago.

But in other ways, bushpigs are also astonish-ingly advanced, demonstrating their genetic diversity by producing no less than four distinct color forms in Uganda alone. And in the forests around Murchison Falls, one of these stylish sounders has made an evolutionary leap that leaves biologists breathless.

They feed on a variety of vegetation, but they are particularly fond of a broad-leafed herb called *Palisota schweinfurthi*— a kind of tropical dayflower. They never disturb the leaves, flowers, or fruits of this plant but dig very carefully down to the tuberlike root and eat just part of it, leaving the rest to regenerate. Somehow, they have learned how to harvest this plant without killing it, practicing what looks very much like deliberate cultivation, husbandry by pigs.

In the end, the only reliable indication of the true evolutionary position of any mammal is its teeth. You *have* to look a gift hog in the mouth.

Early teeth, in the life of individual animals as well as in the history of the entire class, are simple pointy ones with sharp cusps designed to tear flesh apart in the way of their reptilian predecessors. It is only when an animal grows, or its lifestyle changes to take advantage of particular sources of energy, that its dentition becomes more complex. Any change in feeding habits that can increase the efficiency of the teeth offers an advantage to their owner, with the result that mammalian teeth seem always to be in advance of the rest of the body, making them very useful indicators of evolutionary change.

There is an overall plan. The outermost teeth, those at the business end of the snout, become modified for holding and cutting. They are blade or shovel-shaped, have single roots, and seldom number more than three on either side of the upper or lower jaw. These are called *incisors*.

Directly behind the incisors on each side is one larger tearing tooth that functions as a dagger with a very sharp point, or becomes enlarged for display and defense. This also has a single root and grows continuously, requiring open space around it. This, of course, is the *canine*. Beyond it are several teeth, usually no more than four on each side, with very variable form. They can cut or grind, depending on the diet, and are known as *premolars*.

And right at the back of the mouth, where serious chewing takes place, are up to three teeth with multiple roots and broad surfaces. These have grinding cusps and are the *molars*. Every species of mammal has a fixed number of each type of tooth, and the pattern of each set plays a major role in classification and decisions about its relationship to other species. And the way in which each kind of tooth is modified, moved, or even lost altogether, betrays a great deal about how that particular animal lives.

Pigs have the least specialized teeth. An omnivorous diet places no particular premium on any kind of feeding adaptation, so bears and pigs and people have something close to a full mammalian set of forty-two or forty-four teeth, and molars that are low-crowned and round-cusped. And that fact, alone, has caused a certain amount of confusion.

ON FEBRUARY 25, 1922, HAROLD J. COOK, A RANCHER AND CONSULTING GEOLOGIST IN NEBRASKA, wrote a letter that began: "I have had here, for some little time, a molar tooth from the Upper or Hipparion phase of the Snake Creek beds, that very closely approaches the human type." He addressed the letter to the Head of the American Museum of Natural History in New York and offered to send the tooth on if anyone there would care to examine it.

Someone certainly did.

Henry Fairfield Osborn was by then sixty-five years old, nearing the end of a formidable career. He held a chair at Columbia University, served as senior scientist to the US Geological Survey, and was a founder of the New York Zoological Society, president of the museum's Board of Trustees, and a personal friend of President Theodore Roosevelt. He still worked in the Department of Vertebrate Paleontology and was widely respected as an expert on mammalian evolution, with monumental publications on the evolution of titanotheres, dinosaurs, mastodonts, and elephants.

Osborn was a big man, larger than life, with a fondness for big animals and big ideas. He published a book on the *Origin and Evolution of Life,* followed the discovery of Cro-Magnon and Neanderthal fossils in Europe, and was ruefully aware that no trace of such human ancestors had ever been found in the New World. So the possibility of an American hominid greatly excited him.

He accepted the offer with alacrity, and on March 14 received a package containing a single water-worn and eroded cheek tooth. The roots were missing and the crown was heavily worn, but the size and shape and wear were distinctly human. The location of the find, in a deposit about 10 million years old and filled with mammals of mixed American and Asian origin, supported the possibility of a fossil ape coming in on the same Trans-Siberian migration.

Osborn replied immediately to Cook, saying: "The instant your package arrived, I sat down with the tooth, in my window, and I said to myself—it looks 100 percent anthropoid." He suggested that it might be the third upper molar tooth of some higher primate, and he added: "It looks to me as if the first anthropoid ape of America has been found."

On April 25, 1922, in two simultaneous papers in the *Proceedings of the National Academy of Sciences* and the *Novitates of the American Museum of Natural History,* he broke the news: "*Hesperopithecus,* the First Anthropoid Primate Found in America."

This "Ape of the Western World" caused a sensation. Osborn made casts of the lonely tooth and sent them to twenty-six other institutions in Europe and the United States. One of these landed on the desk of the anatomist Grafton Elliot Smith in London. He was enthusiastic too and acknowledged that *Hesperopithecus haroldcookii* represented a third known genus of extinct hominids, along with *Eoanthropus* and *Pithecanthropus.* He published his opinion on June 24, 1922, in the *Illustrated London News,* along with an imaginative drawing showing a pair of very human-looking "Apes of the Western World" with some extinct camelids in the background.

Not everyone was impressed. Arthur Smith Woodward, at the British Museum of Natural History, the original describer of fraudulent "Piltdown Man," was more cautious this time: "The occurrence of a man-like ape among fossils in North America seems so unlikely that good evidence is needed to make it credible." But on the whole, most people accepted that the finding was, as Osborn called it: "One of the greatest surprises in the history of American paleontology."

It was too good to last.

Osborn sent out two new expeditions to Nebraska with the instruction that they were to "run every shovelful of the loose river sand which comprises the deposit through a sieve of mesh fine enough to arrest such small objects as teeth." They worked through the summers of 1925 and 1926, scouring the Snake Creek beds in the company of luminaries such as the great dinosaur collector Barnum Brown. What they found was reported by William King Gregory, Osborn's colleague at the museum: "Among other material, the expedition secured a series of specimens which lead us to suspect that the type specimen of *Hesperopithecus haroldcookii* may be an upper premolar of a species of *Prosthennops*." *Prosthennops* is an extinct form distantly related to modern peccaries. In other words, the "Ape of the Western World" was not human or even a primate. It was a pig.

Those who found Osborn overbearing and patrician had their turn to gloat, but the truth is that molar teeth from most pigs are uncannily similar to their human counterparts. There are details of the cusps on unworn teeth that give the game away, but it is very difficult to tell cheek teeth of older pigs and people apart. The reason is logical, but not obvious.

HERBIVORES HAVE DEVELOPED ALONG ANOTHER THREE DISTINCT LINES. SOME GRAZE, SOME BROWSE, and some pick fruit. Cattle are all grass eaters. Leaf and twig eaters include tapirs and elephants. And the most conspicuous fruit-eaters are the monkeys and apes, though even some of these do occasionally kill and eat small mammals.

Only one group of herbivores eats everything on a regular basis. These are the pigs, whose teeth have now become as generalized and unspecialized as our own, with the result that today there are wild pigs on every continent except Australia—and even there they breed profusely if given the chance to escape from domesticity.

Omnivority certainly increases an animal's choice of habitat and lifestyle, but it doesn't stop there. The stimulus of varied fare has also played a large role in enhancing intellectual activity. Being omnivorous goes hand in hand, hoof in hoof, with being curious, dexterous, and willing to explore new ways of finding, preparing, and keeping food. Omnivores never stop investigating and are always on the lookout for anything in the environment that can be bent to their advantage. They are easy-going nonspecialists, ready to adapt to changing times, and this shows in their brains and behavior.

The cerebral hemisphere in a pig's brain is far larger than the same area in the brain of most herbivores of equivalent body size. There are differences, too, in neocortical development. The proportion of sensory and associative areas in pig brains is far larger than that in strict herbivores, and the result of this seems to be a greater facility for behaving in ways we associate with intelligence.

Pigs are skilled not only in finding a wide selection of suitable foods but also in finding such resources again, even after long intervals. And this sophisticated spatial memory, which plays a large part in foraging strategies, is not easily disturbed by environmental distractions and changes.

And omnivores are more likely to investigate and manipulate novel objects and to spend more time playing when they are young. They are also far quicker to learn new tricks and, almost as important, quicker to abandon new moves that turn out to be counter-productive. In other words, a tendency not to be fussy about your food may be a distinct evolutionary advantage, something that becomes obvious when you look at the history of the largest living pigs.

FOREST HOG

The forest hog, *Hylochoerus*—from the Greek *hule* for "forest" and *khoiros* for "pig"—was unknown to science until 1904.

Olfert Dapper was a Dutch physician who, in his spare time, wrote a travel book about Africa without ever going there. It was published in Holland in 1668 under the title of *Naukerurige Beschryvinge der Afrikaensche Gewesten van Egypten, Barbayen, Libyen, Negrosland, Guinea, Ethiopien en Abyssinie,* and it almost immediately disappeared from view.

But in 1717 it was launched again in Latin with the far more catchy title of *Dapperus Exoticus Curiosus,* and in this guise it proved to be a vital collation of everything known about Africa in the seventeenth century, culled from sources that had since disappeared.

The book describes a West African brightly colored pig called *couja*—now known as the red river hog—and a second much larger and more dangerous black pig with "big sharp teeth, with which it kills and rips open everything that crosses its path." Dapper called this one *couja quinta* and reported that it was to be found in Liberia.

In 1877 Henry Morton Stanley, on his covert and bloodstained transit of tropical Africa, heard of "a gigantic pig two yards in length" said to live in the Ituri Forest of what is now Uganda. And in 1886 the Russo-German explorer Wilhelm Junker returned with the news that a "giant pig" called *nigbwe* was loose in the Sudan.

Zoology ignored all three sources until 1904, when Col. Richard Meinertz-hagen of the British East African Rifles discovered the severely damaged skin of a large black pig in a village on Mount Kenya. He bought it as a trophy and kept an eye open for further specimens. Near Lake Victoria he picked up a huge skull that measured over a yard in length and that had heavy curved tusks and strange lumps under the eyes. And finally, in the forests of Nandi, he managed to shoot a boar that was more than 8 feet in length.

Oldfield Thomas of the Natural History Museum in London christened it *Hylochoerus meinertzhageni* in his honor and described the pig as an archaic and important find, a link between the fossil and the living pigs of Africa. Further finds during the next decade in the forests of Western and Equatorial Africa proved that it was definitely the animal referred to by Dapperus, Stanley, and Junker but not knowingly seen until the soldier tracked it down.

Known now as the forest hog, it can be distinguished by its dark bristly skin, its heavy body, and an astonishingly deep depression between the ears of all males, which looks as though they have run headlong into a cannonball. This "dish" consists of a ring of thickened bone that functions as a buffer in the head-bashing battles between rival boars.

Beneath the eyes are two large, bare "warts" composed of cartilage and glandular tissue that are far lighter in color and that, together with the dish and 6-inch tusk like upper canine teeth, combine to produce a demonic mask that contributes to this pig's aggressive reputation.

I have seen such clashes in the Aberdares, involving rival boars that rushed directly at each other from a distance of 80 feet or more, crashing their foreheads together with a tremendous crunch and repeating the noisy maneuver again and again until one submitted, probably with a headache. The victor then urinated, ground his teeth, and took a bite at the loser's tail as he disappeared.

These pigs live in groups of up to twenty individuals who are most active at dawn and dusk, traveling along well-marked trails, feeding on leaves, roots, fruit, and occasional carrion. They can and will chase hyenas and leopards off kills in their territory. Like all pigs, they wallow for hours and deposit their droppings in designated latrines, which grow to mounds more than 3 feet high. At night, the whole sounder retreats to an elaborate nest, which can cover as much as 500 square feet in deep undergrowth and is accessed only by a long, easily defended passageway. Their young are preyed upon by leopards.

Males are distinctly larger than females, reaching 450 pounds. Sows gestate for 150 days and farrow in special nests made of bamboo stalks, producing an average of four piglets. A sow's piglets will follow her back to the group just a week after birth and may nurse from any of the resident females until weaning at about eight weeks. In times of danger, females give a sharp warning grunt, to which the young respond by lying flat on the ground.

Forest hog feet are broad and separated at the heel to allow the pigs to walk easily on soft ground, while the tip of the snout is very wide and more elastic than that of other wild pigs, suggesting that they cannot furrow or plough extensively but prefer to scrape and browse. If they dig at all, they do so with the lower incisor teeth, which are frequently abraded. The lower canine teeth are razor sharp from contact with the front of the upper canines, and the cheek teeth are well worn from extensive grinding. They have a reduced tooth count of just thirty-four teeth, cutting down on upper incisors and lower premolars.

FOREST HOGS ARE THE largest living wild pigs, sporting a ring of thick bone on the forehead that functions as a buffer in head-bashing battles between the males.

When not clashing thunderously, forest hogs are naturally retiring, working the edges of humid forest up to 10,000 feet and taking refuge in thick cover when disturbed. Their history is one of isolation imposed by a limited menu, and their future is directly linked to the continued existence of such threatened and scattered high-altitude forest retreats.

EVERYTHING WE KNOW ABOUT THE ORIGIN OF PIGS SUGGESTS THAT REAL SWINE, MEMBERS OF THE family Suidae, arose about 40 million years ago in Eurasia in the form of *Propaleochoerus*—the first true pig. But 10 million years later, dry desert and semi-desert areas of North Africa blocked any further movement of northern ancestral pigs into the continent. South of the Sahara, their place in woodland habitats was taken by bushpigs and forest hogs of the two archaic genera *Potamochoerus* and *Hylochoerus* and their immediate ancestors.

Piglike fossils abound in the fifty or more major paleontological sites in Africa and, because teeth survive better than any other body part, these have been intensively studied. In the first half of the twentieth century, various scholars identified seventy-seven species belonging to twenty-three different genera. But in 1977 and 1978, three of the most meticulous field workers got together and, with only the barest minimum of professional bloodshed, culled this confusion into a more workable array of just sixteen species representing seven genera.

There are far more showy and extraordinary mammalian fossils at these sites—rhinos with two-pronged horns, hyenas as big as buffalo, elephants with four tusks, bear dogs and giraffe with antlers, to mention just a few. And yet few of these creatures have received nearly as much attention as the relatively modest and unspecialized pigs. The reason rests on the happy fact that the suids, with their usual omnivorous chutzpah, have evolved and diversified rapidly and enthusiastically, leaving a plentiful supply of good fossils in a wide variety of sites all over Africa during the last 10 million years.

The key to their success as academic subjects, however, lies in their time frame, which coincides with the period in which our own immediate, and far less prolific, ancestors were becoming hominids. Our ancestors' remains have been found in the same deposits as the pigs', and this coincidence has proved very useful in the correlation and relative dating of the two. That works wonders when it comes to freeing funds for further research.

The pigs also benefited from this research, and the picture we now have of their evolution is an interesting one.

In broad terms, pigs have evolved through time in ways that have given them molar teeth that have become progressively longer and higher, and these changes have taken place on a steady basis across all the sites, providing what amounts to a paleontological clock.

Veteran paleontologists Tim White, John Harris, and Basil Cooke have identified four major groups of the fossil pigs involved. Group One contains two large pigs with heavy heads, long tusks, and big differences between the size of the sexes. These are *Nyanzachoerus* ("the pig from Nyanza in Kenya") and *Notochoerus* ("the pig with ridged teeth"), described initially, as is often the case with pigs, from a single molar tooth. Members of this group are the dominant pigs in all early sites between 7 and 4 million years old and accompany the hominids at Laetoli and Omo.

Group Two is confined to the living forest hog *Hylochoerus* and a grab bag of species lumped together as *Kolpochoerus* ("the pig with folded teeth"), an equally large and square-headed species. These are conservative pigs, and they lie on a separate archaic track that leads, more or less directly, to the forest hog. They correlate well with the hominids of the Olduvai Gorge.

Group Three comprises *Metridiochoerus* ("the pig with teeth like seeds"), which is missing most of its premolar teeth, and the living warthog *Phacochoerus,* which shares a reduced set of teeth and eyes that are set very high on the head. All the species in this group are relatively specialized, more likely to be grazers than rooters, indicating a savanna habitat, and are contemporary to hominid finds at Koobi Fora. They take the place of Group One between 3 and 1 million years ago at both late Olduvai and Swartkrans in South Africa.

BECAUSE PIGS ARE OMNIVOROUS, they have teeth very much like our own.

Group Four is graced only with the bushpig *Potamochoerus,* distinguished by its pointed ears, sharp snout, and a lateral flange of bone directly above the tusks in males. Its history is a simple one, passing almost unchanged from 3 million years ago until the present, and it may have originated not in Europe but in Asia. It coexisted with, and may even have traveled with, early *Homo erectus.*

I am not seriously suggesting that the domestication of pigs took place 2 million years ago and was accompanied by another human species, but the "pig clock" that emerges from this work chimes with the notion of a gradual evolution of both pigs and primates during the last 7 million years. We and they were faced with the same environmental changes and challenges, and appear to have responded in the same way, either by lineages that branched, or by ones that held to the same fine line over long periods of time.

It is tempting to see the bushpig and our ancestors at least as fellow-travelers on the same omnivorous path, eyeing each other with caution and a certain amount of curiosity.

ALMOST EVERYONE WHO HAS HAD THE CHANCE TO SPEND TIME WITH PIGS, IN THE WILD OR AMONG domestic animals, is left with a sneaking feeling that pigs are smarter than they have any right to be. They are hoofed animals, after all, and ought to be stampeding mindlessly across the Serengeti, jumping into the jaws of the same crocodiles that lie in wait for them at the same river crossing at the same time every year. Or chewing a cud in some Elysian pasture.

They should not be lying in wait in carefully contrived ambush of real predators such as leopards and jaguars, or standing up on their hind legs at sty doors, looking their husbandmen jauntily in the eye. That's not their job in the scheme of things. They haven't got the tools for it. They lack the essential evolutionary experience of "manipulation." And yet . . .

It is impossible to ignore a pig's snout. It is the first thing you encounter. Pigs lead with their noses, with barrel-like protuberances tipped with a rubbery, mobile disc pierced with a pair of twitching nostrils. This is the pig's advance guard, its chief executive organ and point of contact with the outside world, carrying more nerve ends than any other part of the body. These afferent or incoming nerve fibers send signals directly to the somatosensory cortex of the brain, much of which is dedicated to this source of information.

If humans were to be represented in a way that reflected the importance and complexity of each sensory system, we would look like mannequins with hands the size of beach umbrellas. Pigs depicted in the same way would have snouts as large and conspicuous as snowplows.

Pig snouts are unique and owe much to their habit of rooting in the soil. The tip of the nose is flattened and supported by a tough pad of cartilage designed to allow shoveling into quite hard ground. Two wild boar left idle in the Bronx Zoo took out their tedium by attacking an outdoor concrete run. Beginning with one small crack in the pavement, rooting only with their snouts, they set about excavating and in less than three weeks reduced 4 inches of solid concrete paving to rubble.

The upper edge of the nasal disc is reinforced with bony gristle and raked so that soil is pushed aside by an upward movement of the head. The lower, eating edge runs seamlessly and directly back to the mouth. Nostrils lie in the center of the disc, well placed to collect information and equipped with a net of fine muscles that close up to keep dirt from being inhaled.

The whole *rhinarium*, the nose and everything inside it, is also surprisingly mobile. Don't believe anyone who tries to tell you that pigs lack expression. They are just sparing in their use of such exhibition. In truth, their snouts are finely controlled by a set of three special well-developed muscles, all anchored just in front of the eyes. One broad band pulls the snout up and back; another fine web of fibers controls each side of the snout, exposing teeth when necessary; and yet another powerful muscle strand combines with a strong tendon to pull the snout down. The combination of all three makes the pig's snout better equipped than that of any other ungulate to express a variety of moods and needs, and the whole structure is supported by a bone that exists in no other mammal.

Just in front of the nasal bone and directly behind the disc is the unique rostral bone that adds extra strength to the snout, providing the kind of control in rooting and fighting that the rest of the order cannot emulate. And when it comes to olfactory adaptations, pigs have few rivals. Every approach of a pig's snout carries with it a cloud of information that amounts to a detailed curriculum vitae and that is accompanied by sniffs and snorts that pick up reciprocal items of news.

All in all, pigs are extraordinarily well equipped to lead rich and interesting lives. Their olfactory generosity is largely lost on us and serves only to confirm ill-informed and preconceived opinions that pigs are "smelly." But the notion that they are also "dirty" and "ugly" is absurd. Despite the conditions in which some pigs are kept, and a lifestyle that involves a lot of rooting around in mud, pigs' noses are always soft and pliable, tender, sensual, and moist. And the snout, once you get to know a pig or two, is remarkably expressive and individual, capable it seems of looking beatific, kindly, wise, amused, thrilled, or just plain bored to death.

Warthogs of the genus *Phacochoerus*—from the Greek *phakos* for "warty" and *khoiros* for "pig" —are the most specialized, most ungulate of all pigs.

They are grazers, living on dry, open savanna country to which they have adapted with smaller, sharper hoofs, longer legs, and a faster, more jaunty gait. Their eyesight is good and their eyes are set very high on the head, all the better to see predators approaching through the long grass. Unlike any other wild pig, warthogs have a tail that comes automatically and vertically to attention when they trot away, turning the terminal tuft into a flag that provides members of a sounder with a conspicuous banner to follow.

The head is not only large but also a very peculiar shape. The snout is long and dominated in both sexes by a pair of enormous upper canine tusks that curve outward and backward for up to 2 feet, framing an hourglass-shaped skull wedged between two bony, inverted cones that support the first of three pairs of cartilagenous "warts." There is another pair of unsupported warts between the eye and the tusk, and a third pair in the form of thickenings running along the edge of the lower jaw.

As if this were not insult enough, warthogs are also adorned with sets of long, stiff bristles on their cheeks and brows that make them look a little idiotic, like badly barbered country bumpkins. All of which would seem guaranteed to justify their description as "hideous," were

it not for the fact that the hog itself looks out across this facial wasteland with a bright, almost amused expression in its eyes that turns the whole grotesque effect into a carnival mask worn by a very pretty pig.

THE "WARTS" OF warthogs protect their eyes and snout from damage during tusk-fencing clashes.

There is nothing awkward about a warthog in motion, however. Its barrel-shaped body is sleek and straight-backed, with a dashing mane of long hair thrown casually over the neck and shoulders, tail flag flying high as it picks its feet up like a thoroughbred.

The function of the warts seems to be largely

protective. The first pair cover the tendons of the snout, the second shield the eye, and the third provide armor for the long masseter muscles of the jaw. All these are also well placed to guard against attack from below, the direction from which a slash of a rival's lower incisor is most likely to come, or from the spines and thorns in which the savanna abounds.

There are two species. *Phacochoerus africanus,* the common warthog, is widespread throughout the African savanna, while *Phacochoerus aethiopicus,* the Somali or desert warthog, is now confined to the Horn of Africa, though there once was a population in the Cape Province of South Africa. The species differ mainly in that incisor teeth are completely lacking in the desert form.

The Warthog's dental count is nominally thirty-four, but its grazing specialty has rendered a lot of these teeth redundant, and adults can have as few as twenty-two or even twelve permanent teeth. The incisors are the first to go, then the premolars, and finally the whole grinding function is taken over by one greatly enlarged third molar.

Within their range, warthog boars tend to be larger in the north-—up to 250 pounds in West Africa, down to less than 180 pounds in South Africa. Females are slightly smaller. Both sexes mate at about eighteen months, the gestation period is around 170 days, and an average of three young are born, all with a fringe of white hair on their cheeks.

Warthogs live in small family units and, again unlike their relatives, are largely diurnal. They spend nights and rainy days in burrows, borrowed from aardvark or porcupine, which they often improve and insulate with grass to keep them warm. In areas with high densities of warthogs, "clans" of pigs share a number of burrows where their ranges overlap, with as many as eighteen pigs using the same burrow on any one night. In their core areas, warthogs keep an eye on all likely holes and burrows, checking them out on a regular basis, aware always of the need for a bolt-hole in times of danger. Then they dive down, always in reverse, leaving the door

to each refuge defended, and effectively blocked, by a leathery face full of teeth.

Warthogs graze in the rainy season, preferring short new-growth grasses, often cropping from a kneeling position, cushioned by calluses on the forelimbs. As grasses grow, they are stripped of their seed heads, and in the dry season warthogs move on to exploit the roots and rhizomes of other grasses and small leafy shrubs. When necessary, they will turn to fallen fruit and carrion, and both species are known to excavate and eat mineral-rich soils. They are much less dependent than other pigs on water, but when it is available they drink regularly, wallow, and even swim.

Fights between rival males usually involve frontal clashes of the foreheads and tusks. Warthogs never bite each other. Disputes about dominance include lateral threat displays with manes erect, alternating with breaks for displacement feeding. Submission is signaled by kneeling with ears drawn well back. Courtship involves an arched approach by the male, shuddering like a cockerel in display, until he gets close enough to rest his chin on the female's hind quarters. Her approval is indicated by simply backing into him.

Warthogs in the wild live to about fifteen years. Old animals really do look old. Their straight backs are bent into hollows; their manes become sparse; their warts droop and crack; the skin around their eyes become wrinkled and slack; and their dew claws drag along the ground. There is, however, nearly always a local leopard or lion willing to help bring this loss of dignity to a seemly end.

AS A CHILD ON A FARM IN AFRICA, I WAS LUCKY ENOUGH TO INHERIT AN ORPHAN WARTHOG ABOUT one month old, just starting to feed on solids. Apples, pears, grapes, bread, cake, porridge, dates, chocolates, cheese, and eggs all vanished into his ever-open mouth. He even seemed to like meat, gobbling down chicken and minced beef, but we agreed that it wouldn't be fair to offer him bacon—though I have no doubt he would have downed that just as happily. It was like watching a vacuum cleaner at work, a suction bag with a leg at each corner. That was what we called him: Hoover.

And he grew almost before our eyes. In a month, he doubled his size to 10 pounds, and in three months, at almost 60 pounds, he weighed as much as I did and looked like a proper warthog. His mane began to grow out along his neck and spine, and a pale trim of beard sprouted at the edge of his lower jaw. His warts and permanent teeth began to appear and by the time he was approximately a year old, he stood 2 feet tall and graduated to a more normal diet for a warthog. He grazed the lawn and made long sorties out into the veld around us.

Hoover was a gentleman. It was true that he had a distinctive smell and tended to gobble his food, but in every other way he didn't behave "like a pig." He liked to wallow in a muddy pool on hot days, but he was far more interested in bathing in the spray from a garden hose, or standing under a tap of running water. He was extremely clean and painstakingly polite, never failing to greet every one of us each time we met by nuzzling a knee or, when he could do so, by nose-to-nose or mouth-to-mouth contact.

If Hoover found me sitting or lying down, he would stop a seemly distance away and lean forward until his snout disc was just an inch from my face. Then he would snuffle softly, inviting a response, and if I acknowledged him in any way, he would push his nose right into the side of mine, or into the corner of my mouth, and breathe deeply, inhaling my scent and checking it with his memory of what I should smell like. And if all was well, he would make a happy little sound of greeting and satisfaction.

Pig-ignorant people with fixed ideas labor under the delusion that pigs go around saying "oink" all the time. Some may, but they also have access to a rich repertoire of other vocalizations, and warthogs share a language all their own.

In addition to his greeting sound, Hoover had a contact call, a series of brief, anxious grumbles that let me know where he was when I was out of sight. He also had a louder, more enthusiastic grunt that he used only for the sight or smell of favorite foods. Then there was a soft low call, a querulous sound I only heard when he paused at the entrance to a hole or burrow; a very polite sort of sound that meant: "Anyone at home?"

For more intensive moments, there was an alarm grunt, almost a snort, something sharp and loud that usually preceded flight; and a harsh growling grunt, louder and longer and heard only in threat. This he reserved for all dogs and any strange human visitors.

When first captured, Hoover had produced a sustained long, loud squeal that I have heard many times since from young pigs of several species who are being admonished by their elders, a sound that subsides to a submissive squeal on the point of turning away to flee. Very young warthogs produce a throaty "churr" sound when calling to their mothers, one that changes to a high-pitched "eeek" when they become separated from the litter altogether. And once or twice I have been lucky enough to hear the warthog *chant de coeur* or love call— a loud, rhythmic "putter" that sounds like a two-stroke lawnmower as a courting male approaches his potential mate.

I could recognize at least fifteen different sounds from Hoover. Some were very subtle, scarcely separable to the human ear, but judging by their effect, they carried distinct meaning for other hogs. And I believe I could recognize Hoover's voice and pick him out from other warthogs by sound alone. My grasp of warthog small talk, however, was woefully limited by my inability to detect and decode the olfactory components in his system of communication.

WARTHOGS LIVE ON THE OPEN savanna and need to have eyes high on their heads to help detect predators approaching through the long grass.

Hoover loved to be groomed. He would stand still for ages as long as I kept scratching his ears or ran my fingers through his mane, grooming all the places he couldn't reach himself. Sometimes he would reciprocate by nuzzling my arm or leg in return, but this involved "grazing" nibbles that were far too intense, pulling some of my hair out by the roots. I suspect that my skin was not very satisfying to him anyway, lacking the added interest of all the glandular secretions that warthogs have to spare.

By his second year, Hoover was almost adult, with a fine set of tusks that he polished on tree trunks at the same time as he marked them with his lip and eye glands, leaving a visible stain that lasted for days. He inspected these markers regularly, sniffing at them, refreshing them, adding to their advertisements in the course of a ritual round of inspection that he took at a strut with his tail held stiffly to attention.

FOOD WAS ALWAYS FIRST ON HIS MIND—SHORT GREEN GRASS FOR PREFERENCE, WITH FLOWERS OR seed heads if possible, or failing that, shoots and roots and fruits of a wide variety of plants. I lost count after listing more than thirty different kinds, all plucked and then masticated with his cheek teeth as we walked along. When he had to get right down to it, Hoover would "kneel" to crop more efficiently, resting on his wrists where the bone was already cushioned. Warthogs are born with these calluses, and the fact that they exist before the need for them occurs presents us with yet another of those nice conundrums that life throws up to challenge evolutionary theory.

The next warthog priority is shelter. Since they are the only pigs to live in open habitats, bolt-holes figure prominently in their lives. Fortunately, warthog expansion to the African savanna coincided with the presence there of a large, nocturnal, termite-eating mammal called the aardvark or "earthpig." No relation to the pigs, this bizarre-looking proto-ungulate with asses' ears and a muscular tail uses huge claws to dig down to its prey, leaving many grass-lands looking like battlegrounds, pocked with holes large enough to accommodate a small man or a full-grown warthog.

Aardvark burrows are vital resources in the bushveld too. They go down 3 or 4 feet, usually round a bend or two, and end in a sleeping chamber. Some are quite complex, with several entrances and a maze of passages and caves that provide shelter for porcupines, mongooses, and farrowing warthog sows. Hoover's interest in them began to make sense when at about six months

old he chose to spend his nights away from home, sleeping in a variety of burrows instead of the box with a blanket that we had provided.

Hoover never knowingly passed any burrow entrance without stopping to examine it. He would sniff at it, give the "Anyone at home?" call, then pop down to see what the accommodation was like. Usually it was a quick in-and-out, and sometimes he would pop up triumphantly from a second entrance yards away. On overcast or cool wet days he slept in, coming out only when driven to do so by hunger. Then I used to track him down and wait for him to rise, which he always did in great style, bursting out of the entrance as though shot from a cannon, invisible in his own roiling dust cloud until he was 20 yards away, just in case there was a leopard lying in wait.

Sometimes there was, so he was very careful, coming or going. When choosing a burrow for the night, he always approached it from downwind, coming in cautiously, slightly stiff-legged if he sensed anything untoward, calling ahead before he made his usual acrobatic, backward descent. This was always followed by a quick return, popping back up to the surface for a last look around before he turned in, but on one occasion he went down and just disappeared.

I waited patiently for his reappearance, but when nothing happened I began to feel concerned. Several minutes passed. Then there were some very alarming subterranean sounds and a cloud of dust that billowed up from the burrow as though the roof had fallen in. This was followed by some more scurrying and the welcome sight of Hoover's hindquarters as he backed slowly out of the hole—dragging a 4-foot-long puff adder.

He had the struggling viper by the back of its head and held on tightly as the thick muscular body thrashed away for a while, despite the fact that its back was already broken. Gradually the struggle subsided and I was allowed closer to admire his kill. Then my favorite pig, our alleged grazer, ate every last inch of the reptile.

Wild pigs do kill and eat lizards, snakes, young birds, and small mammals, indeed almost anything they can catch. They learn to do so by example, by being part of a sounder in which such techniques are passed on by adults with the necessary experience. But in the case of snakes with lethal bites, trial-and-error learning is a fatal strategy. Hoover was hand-reared and so lacked such knowledge. Yet, when faced for the first time by a live puff-adder, he was able to use his hoofs to disable, and his jaws to dispatch, an adult adder in style and without injury.

There has to be a very powerful and extensive genetic reservoir of innate answers to a large variety of environmental questions, a blood bank of carnivorous strategies that linger on, even in the species memory of an omnivore that has been adapted to dealing with nothing more challenging than grass. Such programs must include the possibility that there are sometimes snakes in the grass.

Hoover stayed with us until he was three, imposing his personality on everyone on the farm, becoming fully adult, occasionally wayward, more hairy, more warty, and more extravagantly tusked, but never less than courteous and considerate. And he seemed to expect the same behavior of us.

Despite their reputation, most pigs are scrupulously clean. They defecate only in certain places, setting up their "latrines" in agreed spots, even when their surroundings are severely restricted. Warthogs produce small dry droppings that are less odorous than others, but they are nonetheless punctilious about them and exercise a fine restraint. When we walked out together, in company, and Hoover needed to relieve himself, he turned away and went a short distance from us to do so. This may be nothing more than a ritual that ensures that the nest or food is never unnecessarily fouled. But I noticed that, on such occasions, Hoover seemed uncomfortable, dare I say even a little "embarrassed," if we watched.

And that was only one of many ways in which this grizzled and warty, potentially grotesque, wild hog came to seem so delicate and endearing. We lost him, in the end, to a comely young sow in a sounder that passed through our territory on their way north to greener pastures along the Limpopo River. But not before he had taught me some very useful lessons.

NATURAL SELECTION, THE INSTRUMENT OF EVOLUTION, OPERATES ON BIOLOGICAL VARIABILITY, ON the range of new adaptations and ideas thrown up by random mutations in a rich ecosystem. The more novel these are, the better. Most of them are found wanting and are weeded out, but some survive and thrive, adding to the variety and vigor of a healthy ecology.

In any such system there are natural constraints, forces that provide checks and balances, and the oldest and strongest of these is the long-running competition between plants and the herbivores that eat them. As evolving animals develop new ways of exploiting vegetable foods, the plants have had to keep beastly appetites within reasonable bounds by inventing a variety of defensive mechanisms, the most effective of which are chemical weapons.

These can be as simple as tannins that make plants bitter and unpleasant to the taste. Or they can be as complex and subtle as organic compounds that have a delayed action, time-bombs ensuring that insect predators produce fewer eggs or have shorter-lived offspring down through several generations. Such chemical and biological weapons are very effective, but they are also very expensive for plants to maintain. Most plants cannot afford to keep churning them out on the off chance that pests or predators might happen to pass by and drop in for a bite. So what a number of species have done is to develop ways of producing these weapons only when the need arises.

Some acacias and wattles in Africa move tannins into their leaves within minutes of being browsed, and they take twenty-four hours or more to calm down again and return to a state of tasty equilibrium. So kudu and bushbuck have become "gypsy" browsers, feeding for a short time only on any single tree and passing on quickly, like tourists at a buffet lunch.

To ensure that such fast-food cultures are sustained, some plant species such as the hookthorn have set up a chain of outlets. To keep the customers moving, and to ensure that browsers don't stay too long at any one venue, these trees produce pheromones, airborne hormones that they disperse to warn other trees in the immediate area of the approach of a potential predator. These vegetable alarms have the effect of ensuring that the browsers not only keep going but do so for some distance. Some other plants, mainly fungi, have developed very strange bonds with pigs.

TRUFFLE-HUNTING SOWS
respond to a perfect imitation
of the testosterone normally
found only in the saliva of
mature boars.

This begins with a near-perfect partnership between a subterranean mushroom and broad-leafed trees that lack phosphorus and can only get enough of this vital trace element with the help of an otherwise parasitic fungus that lives on their roots several feet below the surface. There are almost 300 species of these "truffles," the best known being *Tuber melanosporum,* which unites with the hairlike rootlets of European oaks to develop symbiotic organs called *mycorrhizae.* These give the fungus access to carbohydrates produced up in the leafy sunlit canopy of the tree, but they return the favor by spreading out in webs through the soil to collect moisture and minerals that are shared with the tree.

The oak reproduces through its acorns, but the fungus is root-bound in the dark unless some other distributor can be found. Some fly larvae and a handful of adventurous rodents do bring a few fungal spores to the surface, but the famous black truffles of Périgord have hit on the preposterous strategy of using wild pigs as their agents. To this end, they have synthesized a perfect chemical copy of 5-alpha-androstol, which is the active testosterone normally found in the salivary glands of boars during the mating season.

It is difficult to imagine the intermediary steps necessary to arrive at such co-adaptation between a mushroom and a mammal, but the fact is that in the dead of winter the truffle produces enough of this pig steroid to filter up through the soil and attract the attention of a passing sow, who digs with huge excitement down to what she believes to be a sexy boar in

a burrow. What she gets is the consolation prize of a delicacy for which gourmets are prepared to pay as much as $1,000 a pound; and what the truffle gets is the benefit of a cloud of spore released by the pig's boisterous excavation.

There is a nice propriety in the fact that the mammal involved in this exchange is a pig. It is hard to imagine the same tidy transaction taking place with a sheep or an anteater. Our acceptance of such creative interdependence in nature requires a more versatile participant on our side of the equation—an omnivore, more like ourselves, or like pigs.

Pigs, in their own way, enter into similar relationships with other plants. They distribute a variety of hybrid fruiting plants that have developed seed coats tough enough to survive digestion. And by rooting in the forest floor, they accelerate the decomposition of organic matter by incorporating forest litter into the soil. There are several studies that show how the presence of wild boars in Western Europe can promote tree growth even in monocultures of conifers on poor soil, and there is good evidence to show that flora in hog-rooted areas everywhere undergoes a change of composition, stimulating a sort of evolution that is not unlike that of human settlement and domestication.

In this light, it is necessary to take a long, hard look at omnivores as the most likely subjects of such selective pressures. The many similarities between pigs and people make them very fertile objects for such close comparison, but first we need to know a lot more about the pigs themselves.

High on the Hog

Pigs are very beautiful animals. Those who think otherwise are those who do not look at anything with their own eyes, but only with other people's eyeglasses.

—G. K. CHESTERTON, *The Uses of Diversity*, 1908

Chesterton was a master of paradox. He used it exuberantly to debunk Victorian pretensions and was quick to exploit the ambiguity of pigs, whose lines he described as "the loveliest and most luxuriant in nature." He was himself cheerfully and unashamedly rotund, insisting that fatness was a valuable commodity, in pigs and in people. "While it creates admiration in onlookers," he said immodestly, "it creates modesty in the possessor."

Chesterton's irreverent essays are a delight to read, largely because he took his comic commentaries to their logical and serious conclusions. He was right about preconceptions and other people's eyes. Some domestic pigs are undeniably, even grossly fat, but their Rubenesque curves are also somehow sumptuous and satisfying, pig potential fully and passionately realized.

As a biologist, however, I am alarmed by the extremes to which artificial selection can lead. In 1933, a Poland China boar in Tennessee was pushed to a world record of 2,552 pounds, the weight of a family car. He was 9 feet long and 5 feet tall, but there is no longer

ARTIFICIAL SELECTION PRODUCES dished faces, twisted teeth, and gross obesity, which is *not* beautiful.

any pleasure, even for the corpulent, in an animal so fat it cannot move without dragging its stomach on the ground.

The boar's name was Big Bill, and his owner, Burford Butler, was very proud of the animal. It won him prizes at every show in which they appeared. He would certainly have seen the pig with different eyes than mine. Big Bill, unfortunately, could never return the compliment, because rolls of fat on his forehead had completely obscured his eyes. This pig was inflated out of existence, losing its identity and its true animal nature in the process of becoming a curiosity.

The art critic John Berger in *About Looking* suggests that humans and animals look at each other "across a narrow abyss of non-comprehension," and that from our side of the divide, the experience has changed over time.

Throughout prehistory, we both hunted and worshiped the animals around us. Our first paint pigments were made of animal blood, and the first subjects to appear in our rock art were the animals themselves. We saw them as both symbol and sustenance—having a friend for dinner.

During most of historic time, animals seem to have been consistently anthropomorphized. We looked at them and saw ourselves, finding courage and compassion, anger and fear, high spirits and low cunning, as it suited our purpose, an exercise in metaphoric zoology.

It is only in the last few centuries that we have chosen to distance ourselves so completely from other species. The industrial revolution removed most of us from direct contact with them. If we saw them at all, it was at a distance, receding into the background of a romantic painting, or framed by the cage of a menagerie.

And the situation in this millennium is not much better. We are entertained and enlightened by superb films of natural history, able to see into every aspect of animals' private lives, but there is no reciprocity, no possibility of interaction. The more we know about them, the more distant they seem to become. Experience teaches us that voyeurs very rarely get the chance to see their subjects looking back.

While the lucky ones among us get to visit nature in the wild, zoos claim to let the rest of us come eye to eye with natural treasures such as mountain gorillas and clouded leopards, but as Berger points out, "Nowhere in a zoo can a stranger encounter the look of an animal. At most, the animal's gaze flickers and passes on. They look sideways. They look blindly beyond. They have been immunized to encounter." There are no surprises, none of the frisson, the emotional thrill that lay at the heart of all the old encounters involving humans and animals looking at one another as equals.

This is a great loss, a withdrawal from the world. We are the poorer for it, which is why I get so excited when an exceptional set of circumstances arises and such eye contact is resumed. It is exciting, but it is also true that, in our desperate search for reconnection with the animal world, we have become too obsessed with our closest relations.

It made sense, of course, to exploit our relationship with other primates. It helps to have so much in common. Studies in the last few decades on gorillas, chimpanzees, orangs, baboons, and vervet monkeys in the wild have produced some fascinating insights. But the most exciting and seminal breakthroughs have been those involving species that do not share our hands, our tools, or our facial expressions, and seem unlikely to feel our need for validation—animals such as bats and elephants and dolphins. Each of these has given us unexpected cause to rethink everything we thought we knew about them, and about ourselves.

And now, I suggest, there is good reason to take a much closer look at some other unusual suspects—pigs, for instance—starting perhaps with the widespread and malicious perception of pigs as "smelly."

PIGS ARE FASTIDIOUS AND SENSITIVE IN EVERY SENSE OF THE WORD. THEY ALL HAVE AN ACUTE SENSE of smell. In truth, they are built for it. The snout is at the same time arm, hand, spade, and primary sense organ: a probe that makes it possible to travel, feed, drink, and interact with others, even in the dark. In tests of acuity, pigs have proven that plastic cards, once nuzzled by them, can be picked out from a pack days later—even after being washed.

The olfactory bulb of pig brains is well developed and is served by an extensive area of sensory cells at the back of the large nasal cavity. This direct chemical connection to the fore-brain deals with the recognition and memory of odors, making it possible for newborn piglets to locate, not just their dam, but the one teat of hers that they have adopted as their own.

In addition to this primary sense of smell, pigs have a functional second system, beginning with the vomeronasal or "Jacobson's Organ" on either side of the nasal septum. This is sensitive to pheromones, airborne hormones that provide information about social and sexual states, and it sends news of these to the limbic system—that part of a pig's hind-brain that deals with the coordination of basic behavior such as sex and aggression. As a result, one sniff of the sexual promise carried on a boar's breath is enough to persuade a sow to fall immediately into *lordosis,* the posture that indicates her readiness to mate, even if she cannot see or hear a male at all.

In return, a female pig urinates more frequently during estrus, leaving traces that excite the boar into standing over and urinating on the same spot, establishing, it seems, some sort of propriety, even in the absence of the female. When the two do meet, they sniff and nuz-zle each other as a prelude to courtship displays that can be expansive and include exaggerated walks, rhythmic grunts, and even a special "mating song." But paramount in all such ceremony, in nearly all species, is the saliva produced by the male as a disinhibitor, which gets splashed around like champagne at a wedding.

From what I have seen among many wild pigs outside the breeding season, the salivary gland is not always so seductive. It also comes into play when males wipe the corners of their mouths on tree stumps and rocks, leaving a visible smear. This appears to be an impor-tant part of general scent-marking behavior involving all face glands on the eye, cheek, and chin in mutual rubbing sessions that sometimes go on for several minutes. Pigs may not defend territories, but they do leave obvious signs that are regularly renewed and become the subject of avid attention from pigs of other sounders who happen to pass that way. And few male pigs confronted with such signs seem able to resist over-marking them with announce-ments of their own.

I worried, when I first encountered all this manic olfactory activity, that pigs were making life far too easy for their predators, but I soon realized that the profit and loss accounts kept by natural selection always balance in the long run. A sounder of any kind of pigs is anyway rather hard to hide, and there are clearly social, sexual, and individual dividends that make such risks well worthwhile.

Mammals are "supersmellers," with the best noses in the business. We not only follow odors; we also manufacture many of our own. We have become scent factories. Our warm blood brews up an extraordinary aromatic chemistry and lets it seep out into the air through every possible aperture in our skins, giving each of us a unique olfactory signature. It is not a secret system. The world in which we live is laced with olfactory information broadcast in ways that make our internal conditions part of the external environment, giving us all a common olfactory connection.

So members of the same litter, sharing many of the same genes, come to have a kind of "hive odor," something that allows members of a sounder to recognize each other at an unconscious level. This creates a chemical cohesion that brings individuals in a mammalian society together into units that are far more closely knit than is possible for amphibians or reptiles. It also allows smell to become a force that prevents inbreeding and encourages individuals to keep some reproductive distance between themselves.

Pig society is smell-bound. Almost everything pigs do is determined in some way by odor. Scent-marking and scent-reading help to define these limits and to cement social communication. In our attempts to make sense of systems that are beyond our sensory grasp, I suspect that we disparage some scent-laying practices by passing them off as "territorial markings." The fact is that pigs are not really territorial at all but operate movable home ranges, shifting these as they adapt to the seasons. They are generous with their secretions for a different reason, one that has more to do with identity than property. They are protecting themselves, rather than their surroundings, finding security in society instead of territory, laying down olfactory perimeters that are flexible.

I feel this way because my first close encounter with a pig was with my pet warthog Hoover, who grew up mostly in human society. He managed well enough, because he had the grounding of just enough pig-time, losing his family when he was almost weaned, to know that his new sounder was somehow incomplete. Hoover accepted our society for what it was, rich in resources but nevertheless imperfect, unpigly. And he discovered that a solitary pig is a sad thing, almost as useless as a bee without a hive. But he made the best of it and came through in the end, I believe, because he had an olfactory safety net.

Hoover's home range was anchored to our farmhouse. This was home base, his resting site. When he outgrew his box and blanket and moved out to a burrow of his own, that loosened the perimeter a little and gave him more territory to patrol, more places to make a mark on—which he did, repossessing focal points each time the busy human traffic overwhelmed and destroyed his signatures. I watched him do it, seeing and sort of understanding what he was doing and why he couldn't rest until it was done. But I also observed something else.

He trotted around on his high heels, tail up and nose down, dashing off after my footprints from the previous day, inspecting new odors, trampling on and trumping them, spreading the word. **AN EARLY NATURALIST'S FAR** too dainty but nevertheless appealing portrayal of a New World peccary.

But every now and then, he would come across his own tracks, and these always slowed him down. They gave him pause for thought. These were different—they were hoggy. They were his! And every time I saw him do this, the same thing happened. They calmed him down. He took time to savor them and they changed his demeanor, giving him a dreamy sort of look, like someone finding solace in the familiar smell of his own armpit. I believe that is exactly how it worked. He was finding himself, encountering the sounder, the hive-smell, the essence of warthog, and learning who he really was.

THIS SENSE OF IDENTITY IS SO IMPORTANT THAT TRUE PIGS, BETWEEN THEM, HAVE DEVELOPED NINE separate glandular areas for the production of an even larger number of odors.

Starting with the feet, there are *digital* glands between the hoofs and *carpal* glands at the wrist on the forelimbs. Moving up, there are *preputial* glands above the penis and *vaginal*

glands on the vulva, and *anal* glands in both sexes. But it is on the face, around the snout, right where a pig meets the world, that there is a concentrated battery of odor-makers. There are *mental* glands on the chin, *salivary* glands in the mouth, and tusk or *buccal* glands at the base of the lower canine teeth. There are *preorbital* glands in front of the eye, and in the corner of each eye an opening for the tear or *Harderian* gland.

All that is missing are active eccrine or apocrine glands in the skin. Pigs don't sweat. For thermoregulation, they rely instead on evaporation, insulation, and special patterns of behavior, such as huddling in the cold and panting or wallowing in the heat. The ideal temperature for most pigs seems to be tropical, around 86° Fahrenheit. Despite their lack of sweat glands, moisture still passes through the skin against the vapor pressure gradient. Unlike us, pigs experience a net water gain under humid conditions, making up for whatever they lose in dry conditions and transforming pigs into animals whose total evaporation rate is very much like our own.

That said, studies on peccaries show that some of these New World animals that survive in very arid conditions do have thicker and larger medullae in their kidneys. They therefore have a greater capacity to concentrate urine and are better adapted to conserve water. Adaptations such as these take time to evolve, but where the peccary is concerned, they had both the time and the space to spare.

Peccaries are a little odd. They are not, strictly speaking, true pigs. Their ancestors lived in Europe 40 million years ago, and 20 million years later some were still to be found in Asia and parts of Africa. In some ways they resembled hippos, having complex stomachs, wide gapes, and canine teeth that grew continuously and remained sharp by grinding against each other in opposing jaws. By about 10 million years ago, the last of them had left the Old World and taken advantage of the Siberian land bridge to colonize North America. We don't know whether they migrated or were forced out by more competitive Old World pigs, but they flourished in the pig-free New World, expanding into South America following the rise of the Central American land bridge about 3 million years ago.

Peccaries are still classified as members of the old suborder Suiformes, but they have been placed in a new family—the Tayassuidae—to distinguish them from the original family of Suidae. *Tayassu* is also the generic name for two of the three living peccaries and derives from the Tupi Indian *taya,* meaning a tarolike corm of the plant genus *Colocasia* that grows on forest floors in the Amazon, hence *tayasu*—"the gnawer of taya."

The common name "peccary" is also Amazonian, from the Tupi-Guarani term *pecari* or *pakira,* describing "an animal that makes many paths through the woods."

Today there are two common peccaries or "path makers." *Tayassu tajacu,* the collared peccary, is small, dark hued, and large-headed, and it thrives in any habitat from tropical rainforest to woodland, savanna, and desert. *Tayassu pecari,* the white-lipped peccary, is taller, heavier, and confined to tropical forests.

Their evolution may have followed different lines, and arguments continue about whether they should be placed in separate genera, but for the moment they are similar enough to be regarded as close relatives. Neither of them has an obvious direct ancestor in the fossil record. Both, however, differ markedly from the third living peccary. *Catagonus wagneri,* the giant or Chacoan peccary, is far larger, with a proportionately big head, long ears, long legs, bushy hair, and a white ruff on the jowl, and it does have direct fossil predecessors in South America who lived almost 2 million years ago.

There is no argument about the differences between Tayassuidae and Suidae, the New World and Old World families. They have been geographically isolated on separate continents for at least 10 million years and effectively segregated by their behavior for twice that long. True pigs have four toes on their hind feet; peccaries have three. Pigs have up to forty-four teeth, including curved canine tusks; peccaries always have exactly thirty-eight, including small, straight canine daggers. Pigs have simple stomachs and a gall bladder; peccaries have stomachs with four extra blind sacs and no gall bladder. Pigs have long tails; peccaries' tails are tiny, with as few as six vertebrae.

This is an impressive list, enough to defend the decision to put them into separate families, but I am even more impressed by their similarities. The two lineages began at the same point with a common ancestor about 40 million years ago, somewhere in the time of the early entelodonts. Their experiences since then have been very different and yet they have gravitated in the same direction. Selective pressures on separate continents have resulted in the same adaptations, producing remarkable convergence. Taxonomists insist that true pigs and peccaries are only superficially similar, but no biologist from another galaxy would hesitate to acknowledge their likeness.

The presence of large heads, long snouts, nose discs, bristly coats, delicate hoofs, an omnivorous diet, and gregarious behavior patterns in both lines is undeniably "piggy." But what clinches the relationship for me is what has happened to peccaries on the olfactory front.

BOTH PIGS AND PECCARIES ARE CONTACT ANIMALS. THEY THRIVE ON COMPANY, TAKING DELIGHT IN their own society, bucking the trend of other forest-living ungulates, such as bushbuck and duiker, which tend to be solitary and shy. The advantages of herding out on the pampas or the savanna are many—more eyes and ears to detect the approach of predators; the "confusion effect" caused by the simultaneous flight of a number of prey animals all fleeing in different directions; the relative safety of individuals in a crowd; and the possibility of defense, or even a counter-attack en masse, against predators. But these factors are less convincing, even canceled out altogether, in the forest, where the density of plant growth presents predators with problems of their own.

I suspect that pigs and peccaries became gregarious in the open and have stuck together under cover not because of the forest's advantages but despite its difficulties. For them, living in a herd, being close to one another, was the most important issue, especially in the forest.

Being omnivorous is not easy. It opens the door to new opportunities and a wider variety of ecological niches, but it also requires a whole range of new techniques. Grazing is an easy option, programmed into every herbivore. But a diet of roots, bulbs, tubers, seeds, fruits, leaves, nuts, grubs, prickly pears, lizards, worms, frogs, toads, snakes, birds, and rats is a good deal more demanding. Collared peccaries can face choices like that in a single week, and they have to deal with finding, catching, collecting, and preparing almost every item in a different way. Such strategies have to be learned, and can only be learned, by living in a society of experienced individuals who already know where to go and what to do when they get there.

True pigs have been faced with the same dilemma, and have opted for the same solution. They stuck together, perfecting contact calls and cementing social bonds, keeping the sounder intact through all the challenges in a million years of evolution, because it made sense to do so. It had survival value. And the glue that held their society in place was the perfumed cloud of pig odor that surrounded them and kept them in touch, through hell and high water, in the Old World.

The gateway to the New World was a cold one. Even during interglacial periods, Alaska was only intermittently tropical. It was accessible only to mammals that could withstand a wide range of temperature and were better dressed for winter. Peccaries were and are.

Unlike many true pigs, peccaries have dark skins and a dense cover of bristles with a unique structure. These stiff hairs have radiating ribs that support a spongy tissue that provides good protection. They are dark-colored and readily absorb heat in winter. In summer, the dark tips of the bristles break off to reveal white faces that give the pelage an altogether less dense and lighter look, reflecting rather than conducting and absorbing heat.

This adaptable coat has allowed peccaries to inhabit a far wider variety of habitats than most true pigs, but it has also become disadvantageous in that less of the body surface is open and available for scent production and dispersion. The usual pig glands are effectively covered up. There are reports instead of a preorbital gland with a slitlike opening just in front of and above the eye in collared peccaries, but this does not seem to exist in all individuals and may be the last vestige of the nine glandular skin areas on true pigs. There is also dramatic evidence of a novel and peculiar structure that appears to have been evolved as an effective olfactory substitute.

When early travelers to the New World caught and killed peccaries for food, they found on the rump, just in front of the tail, an extraordinary button-shaped bump of raised tissue surrounded by bare skin. This looked so much like a second navel, lurking among the bristles on the back, that the German common name for all peccaries is still *Nabelschwein*, the "navel pig." It is, of course, not a navel. It opens up into a deep sac filled with a fatty, highly odorous secretion and is clearly a scent gland with tubular subcutaneous structures leading to a large cistern at the surface. And it has a very curious origin.

The gland is normally not visible, being covered with back bristles several inches long. All the bristles along the spine of a peccary have muscular connections that allow them to be raised in a crest beginning at the neck. Such an erectile mane makes an animal look larger and is typical of most wild pigs, but peccaries have taken this display a step further. The largest of the lower back bristles are arranged in a spray around the gland opening, which is completely exposed when the peccary is involved in full sexual or aggressive behavior.

Lyle Sowls at the University of Arizona has been studying peccaries for forty years and says: "On several occasions I have seen captive peccaries squirt a stream of liquid scent several inches." Usually they do so when confronted by a strange animal. "When first emitted from the scent gland, the liquid has an amber color, but it quickly turns to jet black when exposed to the air." I too have seen and smelled the same dark areas of musky oil on tree stumps and rocks in the Big Bend of Texas. The scent gland appears to function not only as an aggressive display but also as a marker used by both male and female peccaries, and it has become the medium of social communion in all the many daily greeting ceremonies of all three living species, who start mutual grooming by rubbing their heads against each other's rumps, right across the exposed scent gland.

This ritual takes place so often and so freely, between individuals of both sexes and all ages, in captivity and in the wild, that the mutual rub must be regarded as the peccary equivalent of setting up a "hive odor." In other words, it is a common olfactory connection, like that which allows worker bees to recognize one another, or the "colony odor" of carpenter ants, which originates from the queen but grows into a unique olfactory blend that allows each worker to identify another by a split-second sweep of their antennae across each other's backs. It is significant that in peccaries the frequency of mutual rubbing increases when the herd is on edge or alarmed. Then it has a soothing effect, reassuring everyone involved that they are not alone.

On one occasion in Texas, I was able to share this characteristic odor from almost 200 feet away downwind. It is certainly musky, even pungent, but I did not find it at all unpleasant. It struck me as something neighborly and reassuring, a little like being part of a postgame football team in the locker room.

It looks as though peccaries, deprived of the usual multiglandular production of social odors enjoyed by true pigs by having to wear a warm winter coat, have replaced the old pattern with a single new olfactory factory that now serves all the original needs. And they have contrived this by somehow reactivating an ancient avian solution.

Porcupines and some South American rodents such as the agouti also raise long hairs or spines on their backs in alarm, but they have nothing like the peccary rump gland to reveal. The nearest equivalent I know of is the preen gland that in some birds lies right above the tail and secretes an oily substance that is applied by the tip of the bill to feathers to keep them supple and water resistant. This oil produces vitamin D with exposure to sunlight, but even though we are now more aware of an acute sense of smell in some birds, I know of no species that uses scent glands for social purposes—though it might be worth keeping an eye on the oil bird, a kind of nightjar that roosts in numbers in pitch-dark caves in Trinidad and navigates with the help of echolocation.

I am convinced, however, that the warm cloud of friendly odors that surrounds every pig sounder is so important that peccaries, who lead similar lives, have had to go to extraordinary lengths to revive or retrieve something similar. And I feel that their success in this is sufficient to reward them with the status of "Honorary Pigs."

The species name of the collared peccary, *Tayassu tajacu,* comes from the Guarani Indians of Brazil, but elsewhere in the New World it is also known as *chacaro, baquiro, javali, javelina,* and more than a hundred other local variants.

The collared peccary, or javelina, is the smallest and most abundant of the three living species, an all-purpose peccary about three feet in length and 18 inches high. The sexes are very much alike, unless you can get close enough to see the male's scrotum. The rough coat is grizzled, gray-black in color, except for a yellowish neck band or collar that marks the hind edge of a pointed, blade-shaped head. The ears are small and round, and the eyes are distinctly beady. The tail is vestigial, impossible to see, and the barrellike body is supported by short, slender legs. Unlike true pigs, which have two dew claws on each foot, javelina hind feet are different, having only one claw high up on the ankle— giving them a total of three toes against the pig's four.

All in all, a herd of javelina in the open look a little like guinea fowl, always on the move, scurrying along head down or making short rushes at each other. Disturbed, they rush away in tight formation at speeds of over 20 miles per hour. Their average weight is about 30 to 40 pounds.

Collared peccary are the most widely distributed peccaries, from the southwestern region of the United States, through all of Central and most of South America, to the pampas of Argentina and Uruguay. They occupy habitats that range from cactus deserts to high chaparral,

COLLARED PECCARY are the smallest and most abundant suids in the Americas, inhabiting a far wider variety of habitats than most true pigs.

through oak woodland, pine forests, and deciduous brush-land to mesquite bosques, manzanita, and wooded canyons. They seem as much at home on the dry slopes of the Andes up to an altitude of 9,000 feet as they are in tropical forests at sea level. And they handle the long desert passages as easily as they swim large rivers. They are truly hogs for all seasons.

Javelina are physiologically prepared for drought by their efficient kidneys, but most of their other climatic adaptations are behavioral. They have an uncanny ability to find and use microclimates in every habitat. They select shelters and small sun-traps in winter and use caves, tunnels, dense vegetation, and the cool air chutes that flow down valleys during the summer. Historically, they are tropical animals that have modified their behavior in ways that allow them to flourish and find food in any circumstances except snow and ice.

Collared peccaries also show an ability to adapt to human activity. They thrive in secondary growth and agricultural areas, living on crops as long as these offer sufficient cover. In suburban circumstances, they simply become more nocturnal, feeding on vegetables and flow-ers. And where residents are compliant, they come readily to kitchen doors for handouts and show no fear of people. This tolerance alone distinguishes them from the more retiring white-lipped and giant peccaries.

Among themselves, collared peccaries are highly social. Herd sizes range from half a dozen to more than thirty individuals, and they are cohesive. There are no male bands or female harems. Seasonal fighting is rare, and when there are differences to be settled, opponents spar and meet head on, trying to bite each other about the head and neck. Females are generally dominant over males, but there is little evidence of an established hierarchy. Herds tend to feed and rest as a single social unit, and they move slowly across the terrain, tightly knit or widely scattered according to the circumstances. The home range varies from a few acres to over a square mile and may overlap with others, each loosely identified by scent marks along a nominal boundary.

Within the herd, there are a range of interac-tions that can interrupt foraging. Most common of these is a squabble, which usually begins with a staccato chat-tering sound produced by bringing the smooth faces of canine teeth sharply together. This is a threat that can be amplified by the accompaniment of an explosive "woof" with lips retracted to show the teeth. Most often that is all it takes, but if the contention continues, the stakes are raised by locking jaws with one another, whirling around in a wrestling match that ends when one of the opponents is thrown down and has to let go. Submission is signaled by sitting back on the hindquarters, offering a closed mouth to be nuzzled.

In the rare case of more serious fighting, sparring leads to a full-frontal attack in which each animal raises all its back bristles, fully exposing its scent gland as it tries to bite the other's head, neck, or shoulder. This kind of con-frontation is far more noisy, accompanied by raucous growls that can be heard half a mile away and by a braced stance in which the head is raised high and turned to one side to pre-pare for a lateral slash with the exposed canines. Their heavy skull and thick bristles generally shield peccaries from such attacks, but flanks and hindquarters are sometimes scarred. Battles of this kind usually end with the loser kneeling, head down, facing the winner, who deliberately turns and walks dismissively away.

Friendly encounters are aimed at the head, but also at the scent gland on the rump, starting with sniffing and mutual rubbing. But once secretion from the glands has been shared, the two, more often than not, nuzzle each other's flanks and cheeks. And if the encounter involves different sexes, each may sniff at the other's sex organ. It is usually the female who initiates such overtures, which may be followed by courting moves in which the male takes inhibited bites at her neck or shoulder. Then she picks up the pace by riding suggestively up on his rump until he finally gets the idea and they mate.

Gestation lasts for about 150 days, and two or three young are born with tan pelts, a light collar, and a dark spinal stripe. Unlike true pigs, the peccary mother gives birth standing up and immediately tumbles each piglet to disconnect it from the afterbirth. She also suckles standing and at no point makes or uses any kind of nest, though she may take temporary cover in a shelter or a hollow log. The young are precocial, following the mother within an hour or two, communicating with her constantly with a soft "purr" that can only be heard at close quarters, and rushing into a defensive position between her legs when strangers approach. Initially, she protects them against other herd members, but when the whole herd is disturbed she runs away with them, leaving the piglets "frozen," lying quietly, hidden among vegetation or rocks until she returns.

Collared peccaries are very vocal. When a herd disperses, they reassemble by calling to each other with repetitive low grunts. If these fail, they resort to louder, dog-like barks, which are answered until they have all regrouped. When hungry, the young summon their mother back with an irritable complaint that sounds like panic breathing.

Apart from their aggressive grumbles and tooth clacking, adults also produce a loud, sharp squeal in distress, and an explosive series of "woofing" sounds of alarm that accompany every bound as they run away from danger.

Under normal circumstances, it is possible to track a herd of collared peccaries by their background sound—a continuous, busy nasal "grumble," and the syncopated clicking of teeth that accompanies feeding and minor disputes about favorite delicacies.

ALL KNOWN PECCARIES HAVE EXACTLY THIRTY-EIGHT TEETH, A FACT THAT SETS THEM APART FROM ALL true pigs, who sport a range from thirty-four to forty-four teeth, but never thirty-eight.

The families Suidae (pigs) and Tayassuidae (peccaries) have been separated for as much as 30 million years, one confined to the Old World, the other to the New. We know that they had common ancestors 40 million years ago, and we suspect that the peccaries are an offshoot of a line of primitive pigs that went their own way, successfully colonizing both New World continents. In the process, because of similar habits, they and modern Old World pigs have come to look so much alike that they are celebrated in textbooks as a remarkable example of evolutionary parallelism.

Both families have developed long snouts for efficient rooting, and both have greatly enlarged facial muscles to make these snouts effective in an omnivorous way of life. This, in each case, has produced the kind of pointy head we recognize as "piggy" and led to some loose usage of the terms "hog" and "pig" and "swine," which are haphazardly applied to either.

The likeness is remarkable and deserves explanation, but when it comes to a discussion of how and why the two families have converged, it is instructive to look at how the two lineages still differ. A close examination of the heads of pigs and peccaries shows that most of the differences are related to just one development. Everything seems to revolve around the shape and nature of the upper canine teeth.

In peccaries, as in most other mammals, upper canines grow down vertically to meet the lower canines so that the two touch, even when the jaws are open. Such contact results in the sharpening of both upper and lower canines, producing points that are always as sharp as a dagger. Old World pigs, over millions of years, have developed differently. Instead of pointing down, their upper canines grow sideways, at an angle out of the jaw, where they continue to grow, sometimes curving upward with dramatic results—such as the curled 2-foot tusks of the adult male warthog.

Not all true pigs go to such extremes. Bushpigs and wild boar have shorter tusks, but no Old World pig can now use its upper canines as an effective weapon. All peccaries can and do, having interlocking bites as savage as any large carnivore, whereas male pigs seem to have forgone such weaponry in favor of using the canine teeth for visual display. Tusks are certainly effective signs of dominance and rank and, by growing larger and more robust in males, they have opened the door to sexual differences in the size and appearance of all Old World pigs. They are sexually dimorphic. In contrast, peccary sexes are hard to tell apart without very close examination. Peccaries live in herds with sex ratios of 1:1, whereas male pigs gather harems and form family groups with ratios closer to 1:3, relegating subordinate males to bachelor groups or solitary lives.

So one tiny adaptation, the work perhaps of a single gene in one of the two families, produced an initially modest modification that, with the pressure of social selection, changed the whole appearance and population structure of the Suidae.

WITH A LITTLE LUCK, EVERY AREA OF SCIENTIFIC RESEARCH HAS A WIZARD, SOMEONE WHO IS ABLE TO look past the routine, baseline studies and pick up crucial threads that lead directly to the heart of the matter. The clues are usually there, hidden in full view, but it is the special ones who see the big picture first and make it look so easy that their intuitions carry a hint of magic.

Richard Kiltie is one of these. First at Princeton, and more recently at the University of Florida, this New World zoologist has illuminated the study of peccaries with his flair.

Kiltie started in the mid-1970s with a study of the rainforest peccaries in Peru, looking at their interlocking canine teeth and conceding that these do make formidable weapons. But he could not help wondering about the precision with which lower and upper canines meet. It is impressive. They slide into place alongside each other like finely machined parts in an expensive watch. "Why," he asked, "do the opposing teeth need to interlock to such an extreme degree?"

The answer, he suggested, lies in the diet of peccaries. In rainforests, one of the most important and reliable food sources for any omnivore are the seeds of tropical palms. Some of these are large, oil-rich, and rock-hard and can only be crushed with a dead weight of at least 800 pounds. Kiltie calculated that peccaries can exert a force of this order if all the jaw muscles work together, but he also realized that a force of that magnitude would have to be counteracted to prevent the lower jaw dislocating. This, he proposed, is precisely what led to the extraordinary precision of the interlocking canines.

THERE ARE TWO SPECIES OF PECCARY IN SOUTH AMERICAN HUMID TROPICAL FORESTS. ONE IS THE ubiquitous collared peccary, the javelina. The other is a specialist fruit eater, the white-lipped peccary. And, as usual, it was Richard Kiltie who first asked the relevant question: "How do they coexist?"

Kiltie's answer was to look again at what they eat, because different food types often serve as a basis for resource division when two species occur in the same habitat. The two peccaries are superficially similar in structure and anatomy. They have the same shape of skull and similar low-crowned molar teeth, but there are differences in size and use.

Both species crush their food. They are limited to this technique by the shape of their interlocking canine teeth and have only a little ability to move their molar teeth sideways in the mortar-and-pestle method of grinding. Models of their respective jaw actions show that the white-lipped peccary has a maximum bite force 1.3 times greater than that of the collared peccary. This is enough to provide what ecologists call "limiting similarity," a condition that makes coexistence possible.

Kiltie spent three years collecting samples of all the hard nuts and seeds he could find in Peru's Manú National Park, and he measured the force it took to break each of them open in a weight-loading machine. He found that their resistance ranged from 250 pounds to about 2,500 pounds. And when he arranged the results in order of toughness, he discovered that they fell into a very tidy pattern. The top third, which required a load of over 1,200 pounds to break, were too hard for either peccary to crush; the intermediate third, which needed a load of 600 pounds, could be opened and eaten only by white-lipped peccaries; and the weakest third, which gave way under less than 600 pounds of pressure, were eaten by both species.

This division of resources is suggestive and precisely mirrors the contrast between bushpigs and forest hogs in African rain forests, where the smaller pig with its narrower nose digs deeper than the hog can with a broad snout better suited for use as a sort of bulldozer blade.

Not content with parlaying a simple study of the diet of rainforest peccaries in Peru into far-reaching explorations of the nature of adaptive evolution, coexistence, and symbiosis, Richard Kiltie went on to find a link between palm fruits and population dynamics. In 1983 he published a paper with the bland title of "Observations on the behaviour of rain forest peccaries in Peru," barely enough to catch the attention of a casual browser of the *Zoological Record*. But as usual with Kiltie, the hook was in the subtitle: "Why do white-lipped peccaries form herds?"

Why indeed. It is unusual for forest animals of any kind to be highly gregarious, and it is almost unheard of for mammals in the forest to be so. Among ungulates, the usual pattern is for herds to be formed only in open country species. Forest-dwelling species are essentially solitary or found in very small groups. The coexistence of two closely related species of peccaries in the same habitat offered a rare chance to examine and compare their respective adaptations. Kiltie had already shown how the two are effectively divided by something as small as a difference in bite force. Now, finally, he was turning his attention to the most obvious and curious distinction between the two.

Collared peccaries in the forest are seldom, if ever, seen in groups of more than a dozen animals. White-lipped peccaries, on the other hand, are never encountered in groups of fewer than thirty. More usually, they appear in herds 100 or 200 strong, streaming across trails or open areas in closed ranks.

Routine examinations of stomach contents show that both species in the forest are primarily fruit eaters. But direct observation of feeding techniques reveals that their harvesting strategies are quite distinct. Individual collared peccaries pick up nuts and fruits from the surface, barely disturbing the leaf litter, or they focus on deliberate, deep excavations in search of roots and tubers. White-lipped peccaries work in concert, often shoulder-to-shoulder, using their snouts as ploughs, "bulldozing" topsoil, pushing a wave of leaf litter before them, pausing to feed on their favorite palm nuts, which only they can open.

The racket they make while doing this is awesome. White-lipped peccaries are always noisy, barking and squabbling, moaning and retching like football fans on their way home from a lost match. You can hear this hullabaloo a hundred yards away, but the sound of a herd in a palm grove carries much farther. Each crack of a hard nut shell sounds like a pistol shot, and the effect of a hundred fruits being opened and devoured in quick succession has to be heard to be believed. It combines the menace of an approaching forest fire with all the staccato explosions of a Chinese New Year celebration.

Kiltie points out that fruiting palms have a patchy distribution, occurring in groves scattered across the forest, which keeps white-lips from becoming sedentary. They need to travel some distance between feeding sites, and working as a single large group ensures that they get there together and do not find that others have already exploited the patch. But he suggests that the greatest advantage of being gregarious in the forest is the protection it provides against predators. Peccaries are preyed upon by both big cats of South America, puma and jaguar, and there is no possibility of hiding from these if you live and move in a herd that can be heard eating from miles away. Being part of a large group, however, improves individual safety and awareness, and raises the possibility of counter-attacking such predators en masse.

White-lipped peccaries can do that. Instead of running away in a panic, they tend to form a "covered wagon" defense, grouping into a tight circle with a perimeter of sharp teeth. The American ornithologist Stephen Russell found himself facing such a threat in Belize and stood his ground. "But in an instant the herd charged us. We sought refuge on the buttresses of a very large tree. Most of the group veered away but about twenty animals milled below us, some rearing up on their hindlegs to get closer!" Eventually, they dispersed.

TRUE PIGS SUCKLE A DOZEN OR more hairless young, each of which drinks from and defends it chosen nipple against all comers.

President Theodore Roosevelt, in his early hunting days in Brazil, told of a jaguar that stood its ground and suffered the consequences. The big cat felt the force of a dozen interlocking canine-tooth carriers and "was slit into pieces" and eaten.

Hunting dogs today give white-lipped peccaries a wide berth, sensibly refusing to follow them into dense cover, and hunters who do so on horseback have found that they become the hunted, with ferocious peccaries jumping up at them, snapping at their thighs. White-lipped peccaries deserve respect and usually get it.

WHITE·LIPPED PECCARY

The white-lipped peccary—*Tayassu pecari*—is literally "the eater of roots who makes many paths through the forest" and is a dedicated tropical animal.

It is confined to high-rainfall, low-altitude primary forests, wetlands, and palm swamps, and it swims like a fish, routinely crossing all the major rivers of South America. Herds of 400 or more have been seen swimming the River Purus where it is several miles wide and foraging in lagoons along eastern beaches. They enjoy permanent and ephemeral wallows, taking mud baths in the heat of the day.

White-lipped peccaries are taller than the collared species, 4 feet long and standing about 2 feet tall at the shoulder. Their heads are longer and less convex than those of their collared relatives, and the canine teeth in both sexes are more pronounced. The coat is a grizzled black-brown, darker on the head and chest, and the bristles on the lower lips, chin, and throat are cream-colored, as though they have been caught drinking from a milk churn. The ears are a little longer than the javelina's and whitish inside, and the lower legs and hooves are also pale. Both collared and white-lipped peccaries have three toes on their hind feet.

White-lipped peccaries are variously known as *pecari, careto, cachete blanco,* and *cochino salvaje,* depending on their whereabouts, and they range from Costa Rica to the forests of Argentina. They are, however, slow to adapt to changing circumstances and show none of the javelina's resilience when it comes to habitat change or proximity to human developments. In their own habitat, however, they are formidable opponents.

Fruit plays a large part in the diet of white-lips, but they can and do also eat leaves, lizards, eggs, frogs, snails, and insects.

White-lipped peccaries lead hectic social lives. Group cohesion is governed by the gland on the rump, which produces an obviously musky odor as a herd-scent. This is so strong that zoo keepers are loath to include them in any collection, which is just as well because they take very badly to captivity, where no facility can hold enough individuals to form a necessary olfactory quorum. Any fewer than thirty animals is insufficient to provide the kind of society they need, and they rapidly lose the will to live. No one has ever seen a solitary white-lipped peccary in the wild, or a healthy loner in captivity—unless it is one that has been captured very young and brainwashed into believing it is a collared peccary. Then the two might even mate.

In a good herd, white-lips are in constant contact. They probe each other continuously with their snouts, spreading and confirming their chemical identity, touching and nuzzling backs, bellies, head, and groin, inviting mutual grooming, rubbing vigorously across the scent organ. While doing so, they produce a constant sound as well, a sort of long, low, growly rumble almost like the purring of big, bristly cats. This is contagious and spreads rapidly through a group, particularly when they are on the move.

Aggression does occur, usually in the form of face-to-face squabbles between pairs of individuals that can lead to teeth-clicking and sharp barks. If this continues too long, a chase ensues with one of the two hounding the other until it has gone right outside the herd, but this behavior is naturally self-limiting. As soon as the pair realize there is unoccupied space between them and the rest of the group, all aggression grinds to a halt and can even be replaced by nervous "togetherness" calls, which get instant answers from somewhere in the herd, drawing the chaser, and soon afterward the chased, back into the safety of society. No peccary sent to time out can last very long.

The same process operates on a larger scale, ensuring that the herd never becomes too diffuse. Lyle Sowls, the doyen of peccary studies and their management in the wild, has even quantified this, showing that spatial arrangements depend on a "critical distance" of about 6 feet. Aggressive "grumbles" only occur inside that distance. Beyond 6 feet of separation, no intraspecific aggression is possible. Such calls disappear and are very quickly replaced by the "Where are you?" sounds of togetherness.

When a herd is forced to flee, it does so in a tight formation, all headed in the same direction, sometimes

leaving a young male or two as rear-guards. In flight, the herd may add to its vocal repertoire by teeth-clicking as a device to keep contact. This sound travels far and well and is easier to localize in the dark.

The difficulty of close observation of white-lipped peccaries in the wild means that very little is known about their mating behavior. The gestation period is about 160 days, producing a litter of just two young, who are a creamy earth-brown with a dark stripe down their spines and white undersides. They hit the ground running, ready to follow the herd within the hour.

In general terms, it seems that white-lipped peccaries are more raucous in their communication, more demonstrative among themselves, and more aggressive toward outsiders than collared peccaries, and they appear to have even fewer similarities with other ungulates.

WHITE-LIPPED PECCARIES are confined to rainforests, where they form huge herds, routinely swimming across the widest rivers of South America.

IT IS INTERESTING THAT NEITHER OF THE TWO *TAYASSU* PECCARIES GIVES BIRTH TO MORE THAN two young, both ready to stand on their own feet within minutes. This is an obvious adaptation to the mobility expected of a peccary herd. The young are appropriately precocious, like those of most ungulates described as *nidifugous*—"fleeing the nest" or having no nest at all.

Peccaries have only two pairs of mammaries, and often one of these pairs is non-functional. Wild pigs are somewhat more productive and less mobile, producing four or five young in a typical litter. They are equipped for this with six pairs of nipples, or up to eight pairs in some domestic breeds. And in deference to the number of young, these are born earlier and less developed and are *nidiculous,* "clinging to the nest," more like young carnivores.

The mothers of both peccaries and pigs are passive at birth. The rotund little piglets are delivered easily, head or feet first, and usually need no assistance. Mothers seldom clean their young or lick them dry, or cut the umbilical cord, or help to free the piglets from their fetal membranes, but some peccaries will "tumble" a young one if it has become hopelessly entangled. The membranes usually dry and break on their own, and the umbilicus is broken by the piglets themselves as they struggle to get to the nearest teat, which they appear to find by smell and by the pattern of hair on a sow's belly, where all bristles lead to the midline of the udder. Here sows may help a little by uttering short, rhythmic grunts to keep the litter together, and by making nose-to-nose contact with each piglet to set up an olfactory bond between them, imprinting each young pig to her maternal scent.

The young of true pigs are smaller, with a birth weight of just 1 percent of the mother, and far more dependent than those of peccaries. They spend the first days of their lives in a nest in close contact with the sow, who suckles them while lying on her side—an unusual pattern for ungulates, but one they share with hippos and tapirs. This proximity is vital for pigs in temperate areas where the hairless young need to conserve body heat. It is at this time too that piglets produce one of the most peculiar practices of all.

Newborn pigs sometimes start suckling while still attached to the umbilical cord. Within seconds of birth, they try to reach a teat, for preference one of those nearest the mother's head, which usually produce most milk. And in the rush for position, they battle with their litter mates. Once a special teat is selected, the youngster does not drink from any other nipple and defends "its own" teat against all comers, slashing and biting with well-developed canine teeth.

This struggle for precedence takes place within minutes of birth and continues at an intense level, involving very adult-looking aggressive maneuvers, until a teat order is established, usually in less than an hour. Weight gain is directly proportional to the position on the teat line. Piglets born in the second half of the litter, and therefore late to get into the struggle for the "best teats," are twice as likely to die. So, in a very real sense, this struggle for survival actually begins in the womb.

In one experiment with a Berkshire sow, even before her last piglet had been born, eight others were removed from their chosen nipples, mixed up and left a yard away from the udder. All were back on their "correct" teats in less than two minutes. How this suckling order is established, and how the piglet recognizes its selected teat, is still mysterious but probably

depends on smell. Piglets often rub their noses on the udder around the teat in ways that do not seem to be concerned with massaging to obtain more milk. The piglet could be marking the nipple with its own odor, or learning a distinctive smell supplied to each teat by the mother.

PECCARIES HAVE NO MORE THAN two pairs of nipples and usually bear just two young, both ready to run within minutes.

However it works, the aggression involved is curious. It is most unusual for non-human animals to develop weaponry specifically designed for use against their own kind, but this is precisely what pigs have done. The dentition of newborn piglets is surprisingly elaborate and makes no sense for animals that are wholly dependent on milk and won't be weaned for weeks. Most pigs are born with eight fully erupted canine and incisor teeth, not just needle sharp but also angled outward to do the greatest damage in a sideways swipe at a litter mate. And these are not just for show. So many domestic piglets end up with lacerations on their faces that many farmers routinely clip teeth close to the gum at birth.

This situation is so extraordinary that it even prompted David Fraser, a Canadian agricultural scientist publishing in a learned journal, to submit the summary of his paper on "Armed sibling rivalry among suckling piglets" in the form of a slightly puzzled poem:

A piglet's most precious possession
        Is the teat that he fattens his flesh on.
He fights for his teat with tenacity
        Against any sibling's audacity.
          The piglet, to arm for this mission,
Is born with a warlike dentition
        Of eight tiny tusks, sharp as sabers,
Which help in impressing the neighbors;
        But to render these weapons less harrowing,
Most farmers remove them at farrowing.
          We studied pig sisters and brothers
When some had their teeth, but not others.
        We found that when siblings aren't many,
The weapons help little if any,
        But when there are many per litter,
The teeth help their owners grow fitter.
          But how did selection begin
        To make weapons to use against kin?

It is true that the battles are brief and soon subside once teat order is firmly established, but such precocious aggression, coupled with the necessary weapons and motor skills, is alarming. It appears to be the adaptive result of an extraordinary arms race stimulated by sibling competition. And the presence of fully erupted teeth at birth does not seem to be an aberrant product of domestication and larger litters. The same dentition is found in wild boars and peccaries.

The loss of a teat in the first few hours of neonatal competition is not fatal—a beaten piglet simply moves down the line to the next available nipple. But a loss as late as the following day can be lethal. By then, it is too late to compete, and such losers often weaken rapidly and die.

Nursing is not continuous. Sows respond to persistent prods and demands by "letting down" milk from time to time on a cyclical basis about once every hour, so some piglets maximize their consumption by falling asleep between meals with the nipple still in their mouths. Everyone gets dislodged when the sow stands up to feed herself or rearrange the bedding with her snout. When she is ready to return, she roots all the young into a huddle so that none gets squashed as she lies down again, but in cramped domestic pens 10 or 15 percent of young die as a result of being sat on.

There are obvious advantages in staying close to mother and getting a good share of her favors, but there has to be a trade-off between feeding and the risk of being crushed when she suddenly lies down or rolls over. The piglets who are not gaining weight properly usually take the greatest risks by constantly nuzzling the udder of a standing sow, but nothing goes to waste. Mothers usually eat their dead infants.

BEFORE LONG, THE SOW ESTABLISHES A ROUTINE, INVITING HER PIGLETS TO SUCKLE WHEN SHE IS ready, calling to them with a snoring sound. She continues the call intermittently throughout a feeding session. When she stops, milk stops and everyone takes a nap, which continues until some of the more demanding young put out a plaintive squealing sound. This is very different from the high-pitched scream that all piglets of all species produce when isolated, or sat upon, bitten, or handled. That is an unmistakable call for help and gets very quick attention from the sow.

The mother has two alarm calls of her own. One is more of a warning. It is low-pitched and repeated, a "Now hear this!" announcement that she is about to move, and it continues right through her action, like the insistent beep of a delivery truck reversing. It seems to be a mild version of the threat grunt that warns anyone getting within a pig's minimum distance, but it is modulated to suit the circumstances. When hearing it, the piglets back off and move a few feet away, huddling together as they wait for the next development. If anyone doesn't get the message, the sow nudges them away or even picks an offending piglet up in her mouth and drops it impatiently out of harm's way.

The second adult alarm call is a far sharper and more distinctive grunt. It is abrupt and explosive rather than vocal, almost a sneeze. On hearing it, all adult animals stop what they are doing and look at the source of the sound. The piglets drop to the ground and crouch motionless until the alarm is over. The sow assesses the danger, sometimes clicking her teeth to keep the herd alert, and she may take the initiative herself, leading a group of adults in a terrifying concerted pack attack. The younger the litter, the more ready the mother is to defend them. As they grow older, her readiness and their need for protection slowly fade away.

Among true pigs, a sow usually keeps her young litter in the nest until they are walking on their own. Then she introduces them to the herd, starting with other females. The kind of association this provides depends on the species, but in some cases assimilation into the sounder takes place quite quickly. Among warthogs, for instance, there may be a certain amount of *allosuckling* in which young are allowed to nurse from another mother.

The evolution of such behavior offers a challenge to evolutionary biologists. Lactation is a costly business for a female and the resource she provides ought, theoretically, to be reserved for *autosuckling*—feeding her own young, who share half her genes. But suckling away from home happens often enough to suggest that there have to be good reasons, high benefits of some kind, to make it all worthwhile.

Studies show that allosuckling is not a case of mistaken identity or milk theft. The sows involved clearly recognize their own young and can distinguish these from other litters, and yet they still allow it to happen. The benefit for alien piglets is obvious—they enjoy the fruits of two wombs—but the advantage for other mothers is harder to explain. They are capable of nursing selectively and yet choose, on occasion, not to do so.

Fallow deer are reciprocal nursers. In a big herd, does whose young receive donations of milk from other mothers are themselves milk donors. Everyone wins. But in groups of wild warthogs, the frequency and fairness of eating out are not so tidy. A study in Uganda showed that juveniles that drink from more than one mother do manage to get more milk than their less larcenous litter mates, but it also showed that allosuckling most often takes place from those mothers who have smaller litters and can afford to be generous.

None of this, however, explains why it happens at all. Altruism is defined as "self-destructive behavior performed for the benefit of others," and that goes against all genetic imperatives. Natural selection should not favor animals who improve the fitness of others at their own expense, at least not in the short term. In the long term, however, the costs and benefits require a different kind of accounting. Genes for parental care tend to get passed on eventually to offspring who, in turn, care for their own young. In the long run, the result is a population in which the tendency to care is relatively more common than the tendency to not care, and the result is a society in which individuals tend to be more generous than selfish. So the interests of the sounder come first, and in some species such unthinking benevolence becomes encoded in the genes.

In a collared peccary herd, only two or three adult females are regularly impregnated and bear young. The young are assimilated into the sounder within days, but for some weeks, the only other individuals allowed near them are young female relatives of the mother. These are often her younger sisters, and their first task is to eat her hormone-rich placenta at the birth. They are effectively inoculated in this way and become peccarine "nursemaids." The piglets quickly become accustomed to them and include them in their quest for milk. They thrust against the udders of their adolescent maiden aunts and eventually this stimulation results in lactation that makes it possible for the nursemaids to join in suckling their adopted nephews and nieces. This happens even though they may be only six months old and are still being nursed by their own nursemaids in an extraordinary cycle of the sounder's estrogens.

The biological function of these nursemaids is clear. They assist a mother in feeding her young even when the herd is on the move, and they stand ready to take her place if she dies. And the fact that they are being suckled prevents them from becoming pregnant themselves until they are fully mature. In the meantime, they are gainfully employed in assuring the survival of young carrying some of their own genes, while they gain all the experience necessary to become full-blown females in their own right.

THREE OR FOUR WEEKS AFTER BIRTH, MOST PIGS AND PECCARIES ARE READY TO TRY SOLID FOOD. They become curious, and the largest and strongest in the litter begin to venture away from the sow, keeping up the contact call to bolster their own courage and to let her know where they are, dashing back at the slightest hint of anything untoward. But, in truth, it is the sow who decides when they should be weaned.

As soon as the litter has grown big enough to force a sow to stand and deliver, by massaging her sides and belly to trigger the sucking reflex, she begins to put some distance between herself and the hungry little mouths. She leaves her young for longer and longer periods, giving herself a break and getting them used to the idea that she is not a fixture in their lives.

A sow will generally take time off during the day, choosing to return to the young only at night, but each new day she makes greater efforts to get them to follow her, until they are ready, as a litter, to leave the nest altogether. This exodus is accompanied by a gradual reduction of milk yield and more and more encouragement to seek out and try other sources of nourishment, but full weaning in some cases can take as much as sixteen weeks. Any attempt to shorten this process in domesticity results in stress for both mother and young and can lead to weight loss and lethargy among the weaners, who then show all the signs of acute depression.

Lyle Sowls, "The Peccary Man," who has made extensive studies of their management in the wild, points out that the collared species, in most of its extensive range, has had to adapt to periods of abundance that alternate with prolonged periods of scarcity. These are governed largely by seasonal rains, which have a direct effect on the production and survival of the young. There is a marked reduction in the numbers of young in dry years. The exact cause of this is still unknown, but Sowls suggests that the mother peccaries may exercise their own kind of population control. Litters still appear at the expected times, but sows eat their own young—sometimes at birth but also later in the weaning if it is accompanied by severe drought.

Collared peccaries have learned how to survive drought by taking advantage of the moisture in succulent plants. In most of Central and North America, this means cactus defended by a wide variety of very sharp spines and toxic chemistries. Peccaries have found ways of dealing with these if they must, using their front feet to hold prickly pear pads down against the ground, peeling off the skin that carries the spines on one side and eating only the soft juicy pulp within.

They also take care to choose those species that have the fewest spines and the lowest levels of chemical deterrents.

Sowls tested one female peccary kept in captivity and fed on the usual commercial pig feed. She had already successfully reared three normal litters under these conditions, but when fed only on prickly pear cactus in the sort of quantity that would normally be available to a wild peccary under drought conditions, she changed her behavior. Her gestation ran the usual length and her fourth litter of twins was born on the expected date, but she immediately ate them.

Mortality rates in most piglets are highest during the first forty-eight hours, when more than half of all preweaning deaths occur. Newborn pigs and peccaries have poorly developed immune systems, and their lack of subcutaneous fat makes them highly sensitive to cold. So natural deaths do occur and are disposed of by the sow, who eats them, just as she ingests all other foreign matter in the nest area. Infanticide is also not uncommon, in captivity or the wild. One study records thirty instances in a captive herd, all in the first twenty-four hours after birth. But in almost all such cases, the culprit is an unrelated adult attacking an unguarded juvenile. Cannibalism is usually regarded as an artifact of captive conditions in which unrelated or unacquainted animals are forced into unnatural proximity. But little is known of exactly what happens to piglets in the wild: Eating one's own under harsh environmental conditions may be an act of necessary and unselfish kindness.

WHILE INVOLVED IN AN ARCHAEOLOGICAL SURVEY OF THE SONORAN DESERT IN NORTHERN MEXICO, I made my first nose-to-nose contact with a young peccary.

He was the last survivor of a hunt with dogs that killed or scattered a small herd of collared peccaries and would have accounted for him too, had we not heard his loud, squealing distress call. We followed it for half a mile or more, finally discovering the desperate little piglet lying motionless among rocks in a dry arroyo where his tan coat blended in perfectly with the background. That wouldn't have saved him for long from coyotes or bobcats after dark, so I carried him back to the estancia where I was staying and took over where his mother had left off.

Within a month, "Salsa"—so named for his zesty personality—was a part of the family. He had a very lively interest in absolutely everything, rushing off in all directions and stopping halfway as soon as something else caught his attention. He had an inquiring mind, limited by an attention span too short to satisfy his curiosity, so life often left him squealing loudly with frustrated indignation. So much to do, and so little time . . .

Until he got stuck in a drainpipe. He was hot on the trail of a rock lizard, one of the craggy ones that did push-ups on the stone walls of the patio. The lizard stood its ground until he was just a yard away, then skittered off down a path that led into a drain between one terrace and another. Salsa pursued it into the pipe in his headlong rush toward daylight at the other end, but he never got that far. The pipe narrowed and, halfway in, his head and shoulders jammed, leaving his hind feet scrabbling helplessly behind him. He let out a terrible scream of alarm and anger, not quite understanding what held him in such a vicelike grip and quite incapable of getting enough purchase to back out again.

Fortunately I wasn't far away, but the sight of me at the safe end of the trap only made him scream twice as loudly and struggle further into trouble. I went to the barn for help and found a plumbing tool, a set of bamboo poles that screwed into each other and ended in a large rubber disc that fitted the drainpipe perfectly. I pushed this plunger slowly through until I could feel Salsa struggling against it, then gently applied enough pressure to push the little peccary backward away from the constriction until he was free. He popped out, gave himself a shake, and trotted off as though nothing had happened.

But it had. From that moment on, Salsa was a different hog. He still had an inquiring mind, but his curiosity was tempered now with caution. He was as nosy as ever, but he had been frightened sufficiently to think twice before rushing into such a situation again. It slowed him down just enough to engage his brain in each new venture, and he soon learned that it was worth dealing with one thing at a time. In this way, he discovered the possibilities that opened up with every new opportunity, and I could see the intelligence growing in his eyes.

Peccaries and pigs get used to pushing things around with their snouts, proboscillating the world around them, picking up objects in their mouths, throwing them high in the air. Salsa soon began to collect things that pleased him and to bring the best ones home as trophies of the hunt. He showed them to me proudly but was loath to let go, and when one day I tried to take from him what looked like a loaded shotgun cartridge, he discovered play.

Salsa ran a short distance away and I pursued him. He stopped, letting me almost catch him, and then dashed off again, and we were involved in a chase. At one point, he dropped the prize and I got to hold it long enough to find that it was a safe, spent cartridge. So I ran away with it, and he chased me. And very soon we were involved in social play, in the age-old "play-fleeing" game in which roles are rapidly reversed, letting the pursuer become the pursued, going on and on until everyone involved is exhausted and happy.

That game opened a window for Salsa. He had never played before but was soon an inveterate player, initiating games with everyone, bounding like a puppy with his head on one side, panting in open-mouthed invitation, looking around for an object, any object we could pretend was worth a chase. And if we were too tired or too busy, he seemed perfectly happy to go away and play on his own, chasing his nonexistent tail, whirling around in circles, trying to scratch his head with his hind foot, jumping up and down on the spot.

Jumping where there is nothing to jump over, running without going anywhere, fleeing when there is no enemy to flee from—all these are actions that lack any obvious function. They appear to be undertaken purely for pleasure and are apparently not performed by invertebrates or fish or amphibians. They are patterns peculiar to birds and mammals and are most often initiated by animals that spontaneously seek out new situations on their own initiative. Which means omnivores, of course, and especially pigs. Play is epidemic among young pigs.

We call such behavior "play" and find no difficulty in recognizing it when we see it. It is easy to distinguish. An animal involved in play-fleeing or play-fighting looks very different from one seriously occupied in flight or fight, but it is very difficult to define that difference.

Play has no goal and is not confined to any particular activity. Anything an animal does in earnest can be done in play. Young peccaries play chase, play flee, play fight, play leap, and play wallow. They do all these things as part of coordinated herd activity, but they tend to put away such childish things after they are more than four months old.

But it would be wrong to regard play just as something opposed to work. It is far more important than that.

Play is voluntary. You can't *make* someone play or legislate play into being. A pig wearing a silly hat and jumping through a hoop isn't playing. Play implies pleasure, fun, and a definite lack of constraint, something that comes more naturally to the young than it does to adults.

Play is paradoxical. It occurs at inappropriate times and is often contradictory, seldom putting objects to their normal use. Young wild boars chase windfall apples as readily as kittens chase balls of wool. A lot of playtime involves friendly fighting and inhibited biting, and the most serious adult pigs can sometimes be caught frisking, leaping, and kicking up their heels for no reason at all.

Play is random, repetitive, and ritualized. It borrows from, and mixes up, a wide variety of adult behavior, and then repeats parts of these endlessly, or subjects them to new and arbitrary rules. I have watched young warthogs chase each other for possession of a white pebble that can be wrestled away from its temporary owner at any time, but not if he or she is standing on a particular low anthill that has, by general agreement, become a safe haven.

Play is stimulated by novelty. Omnivores cannot resist anything new and different. Anything strange in a pig's environment is immediately examined, sniffed at, and snouted, and if it is not found to be edible, incorporated into some kind of game. Hoover, my warthog, could sometimes be seen trotting around with a favorite plaything, an old tin cup, balanced jauntily on the tip of one tusk, where it made a cheerful sound. And I know of a hand-reared bushpig that never retired to its chosen nest without a pink rubber toy pig for company.

Play remains elusive. It is almost certainly a complex collection of activities that are not just frivolous. The amount of time spent on it by young animals suggests that it is important; and a lack of it may impair the orderly acquisition of several vital social abilities. Play seems to be necessary for a healthy brain in pigs as well as people.

WHEN IT COMES TO COMMUNICATION BY SOUND, PIGS ARE PROVIDING SIMILAR SURPRISES. THEY have excellent hearing, with a frequency range much like our own, though this extends further into the ultrasound band up to about 40,000 cycles per second. We cannot hear anything above 20,000 cycles per second. When it comes to ear size and sensitivity, there is no contest. Pigs have huge pricked ears that act like radar dishes tracking moving sounds, even when they are asleep. They also have a sound localization threshold of just four degrees, which puts them up with the most accurate localizers of sound in the animal kingdom. The pig response is closer to that of cats than it is to other ungulates. If they have any weak acoustic spots, these are in amplitude. Pigs hate sudden loud sounds. All noise pushes up their heart rate and blood pressure.

As recently as 1972, it was believed that pigs produced no meaningful sounds or discrete acoustic signals, and that their vocalizations fitted on to a continuum that rose and fell only as an indicator of their general state of arousal. A decade later, a study of the voices of suckling pigs produced five different classes of sound—croaks, deep grunts, high grunts, screams, and squeaks. Now, it is grudgingly admitted that pigs are not only extremely vocal animals but also capable of "honest signaling," sending out information that conveys news of their true state and condition.

This dry conclusion masks an argument between agricultural scientists, who manipulate domestic pigs in the interest of productivity, and field naturalists who follow the herds, recording calls produced by wild pigs and peccaries in societies faced with a wide variety of environmental circumstances and conditions. Out there, the air is thick with the sound of sounders in full cry.

Just for a start, newborn piglets learn to recognize and respond to their mother's voice within a few minutes. Before they are an hour old, the young can distinguish between their own mother and other sows by sound alone. They respond to, reply to, and move toward playbacks of her call to the exclusion of all others. They may even have become familiar with her voice while still in the womb.

The replies of the piglets are also meaningful, sending information that is not only directional but also informative. Piglets isolated at a temperature of 56° Fahrenheit use higher frequencies and longer calls than litter mates kept at a comfortable 86°. They communicate their need for both food and heat, telling of hunger and discomfort, conveying an urgent need for resources other than milk, to no one but their mother. These are unquestionably reliable signals of real need that produce appropriate responses only from the sow in question, who stops letting-down and gets up to find the source of distress.

Later in life, the range and function of calls increases well beyond simple information about the availability of milk or nutritional need. Pigs vocalize to attract or ward off litter-mates; to announce the discovery of food sources or water; to advertise sexual readiness and alarm; to address fellows; and to intimidate rivals. Or, and this is an equally important signal, they sometimes choose to be absolutely silent. Saying nothing can often be a very useful strategy.

Our understanding of pig talk is still very limited. At a fundamental level, we know that isolated domestic pigs respond with grunts and squeals. In general, short single grunts accompany investigation of any novel stimulus: "What is this?" Longer single grunts at a higher level of arousal act as contact calls: "I'm over here!" Squeals of all kinds are intensifiers, adding urgency to a communication as required: "Come quickly!"

Nonvocal sounds such as huffing and puffing, which disturb the air and make volatile odors more available, also add a measure of interest to the signal. All such messages become more frequent and more focused when there is another pig, or even a human, nearby. But, for the moment at least, this conversation is very one-sided. The way in which all these messages are pitched and shaded, creating subtle differences in threats and warnings, contacts and greetings, is still largely lost on us.

Lyle Sowls has made a very brave attempt to classify the sounds of collared peccaries, distinguishing between "togetherness" and "aggression or alarm" calls.

"Togetherness" includes everything that brings individual animals together and keeps them in close contact. Loud "barks" alert herd members to the need for regrouping, and "low grunts" accompany the process of reassembly. Once reassembled, the sounder is made more cohesive by repetitive short, low-pitched sounds that confirm contact between adults. In the young, this manifests itself as "purring," which can become "complaining" if any infant feels insecure.

"Aggression" is conveyed by a sliding scale of repetitive sounds, beginning with "grumbles" and "growls" that accompany feeding and any other competitive behavior that controls spacing or confirms the order of dominance in the herd; and going on to the more direct individual challenge of "tooth clicking." But when these go beyond normal control, they become loud squeals of submission or distress, both of which are overruled by the explosive "alert call" that puts herds to flight.

All true pigs share a similar repertoire of grunts, squeals, and snorts that express well-being, arousal, rage, fear, and pain, but the best analyses to date are those on warthogs by Hans Frädrich in Germany and D. H. M. Cumming in what was Rhodesia. Both naturalists divide pig calls into three groups: "grunts," "growls," and "squeals."

"Grunts" range from the "soft low call" that keeps a litter together, to the more abrupt and boisterous sound of a "greeting call" between adults. This is most intense between individuals that have been temporarily separated, and becomes a more plaintive "location call" when two hogs can hear but not see one another. And at the emotional peak of all possible grunts is the splendid *chant de coeur* or "love call," a loud rhythmic attention-getter produced by courting boars.

Grunts become "growls" in situations that involve threats and alarms and range from the warning "wooomph," a low-pitched expulsion of air, to the full explosive bark or "alarm grunt" that needs to be taken seriously and usually leads to rapid flight.

"Squeals" are more difficult to define and depend far more on context. They run from little "submissive squeals" that accompany defeat to "squealing growls" that are more throaty and drawn out in conflicts between two well-matched opponents. And then there are "discomfort squeaks" that are confined to hoglets who feel put upon, and "long loud squeaks" by individuals of all ages who panic when they are attacked or handled.

Every species of pig or peccary has its own variations on these themes, depending on their herd size, habitat, or history. But broadly speaking they share an acoustic repertoire that cuts across taxonomy and makes it possible for all members of the suborder to have a general understanding of each other's vocal signals. Warthogs and bushpigs respond to the same kind of loud snorted alarm. Collared and white-lipped peccaries in forested areas use very similar low-pitched "grumbles" to keep contact with one another in dense cover. And there are also species-specific niceties that give each kind of pig a finely tuned system of communication that amounts almost to a language.

MY TAME WARTHOG HOOVER HAD ONLY A VERY BRIEF EXPERIENCE OF WARTHOG SOCIETY, BUT IT WAS enough to have taught him, or to have released in him, a working knowledge of warthog communication. As he grew, he used most of the acoustic signals I later came to recognize in the wild. He greeted and threatened and submitted with all the appropriate sounds, behaving when he walked out with me as though he was part of a bachelor herd. And I encouraged this, replying in kind with my best approximations of warthog small talk.

This sometimes worked and sometimes left him looking at me rather quizzically, wondering what on earth I was trying to say, but on the whole it helped. I grunted and growled in most of the right places, and he granted me the benefit of the doubt and the very real gift of his own superior senses. He always heard sounds before I did, smelled odors before I could, and somehow picked up news on his early warning system well before I would.

On one occasion, Hoover set off his short sharp grunt of alarm and went to ground when a martial eagle flew by and let its raptorial shadow fall on our path. But I noticed that he responded differently to dogs, snakes, and human strangers. All these potential threats were announced with the same "alarm grunt" but were preceded by a caveat, a qualifying clause in the form of a warning soft snort or "woomph," which modified the alarm in a way that allowed Hoover to stand his ground. It seemed to say: "We have a problem, but there is no need to panic yet."

So I didn't. But it wasn't until twenty years later, when I was completing my doctorate in ethology, that I realized what Hoover had done. Linguists decree that true language involves a set of rules known as syntax, in which word order is linked to, and changes, meaning; something only humans can do. But it seemed to me, in retrospect, that Hoover had done exactly that. He had shown that the "alarm grunt" was a matter of life and death, an announcement to be taken seriously, and meant "Run away!" Then, by linking it to an extra sound in the sequence "woomph warning + alarm grunt," he had softened the call so that it meant: "Get ready to run away!"—which is very different.

Ever since then I have looked hard at warthogs and listened to their chitchat, but it is difficult in the wild to recreate the intimate connection I had with Hoover. Our closeness had made it possible to pick up shades of meaning that are rather subtle and easily missed. I haven't heard that precise "woomph + grunt" again. It might have been an aberration of our particular interspecific communication. Hoover could have been making allowances for the snort-challenged humans around him, but I suspect there was something more going on. I think that pigs everywhere are dropping semantic pearls before human swine who labor under the delusion that all pigs can do is go "oink."

A good test of such communication in the wild would be to look more closely at some of the relationships that already exist between pigs and other species. Bushpigs in Africa follow baboons and vervet monkeys, bearded pigs in Indonesia follow gibbons and macaques, all feeding on trees in fruit and enjoying mutually increased protection against predators by the addition of extra eyes and ears. I predict that such situations will be found to feature interspecific signaling even if it only involves shared alarms, and that the vocabularies involved will be found to employ forms of syntax that save lives. And I suspect that some pigs, among themselves, are already putting such sophisticated signals to work in aid of more efficient cooperation.

use their alarm calls more freely and more frequently when foraging anywhere near any of the small herds of a larger peccary known to locals only as *tagua*. Pigs evolved as nocturnal forest dwellers, so it was inevitable that their senses of sound and smell would predominate, but their sight also plays an important role.

Most sources still dismiss sight in pigs as "poor," even "negligible," but that has not been my experience. Hoover could see me coming, even from upwind, from over 200 yards away and would come running to greet me, tail tuft held high. Tests with domestic pigs have shown that they can discriminate between people at a distance by sight alone, and they learn to do so very quickly, even when test subjects wear identical clothes. They react, it seems, to body size, shape, and facial features.

Pigs' eyes are the same size as ours and have similar ocular power and color vision. Pigs' retinas are rich in short, thick cone cells. Their pupils and lenses are very similar to ours and their binocular field is about 12°, compared with our value of 1.4°, but we both belong near the same end of the scale compared to 130° for cows. Pigs also have relatively wide peripheral vision, approximately 310°, giving them an overall panoramic view of their surroundings. All of this suggests that conclusions such as "visual communication is not considered to be of great importance in the pig" are very shortsighted.

Vision was of sufficient survival value in the evolution of the warthog to reshape its entire skull in order to bring the eyes into a unique position high up and far back on the head. To achieve this, the lachrymal bones had to be enormously elongated and the orbit lifted in a way that puts the eyes up on the crown of the skull, like a pig with a periscope. The result is that the warthog, which is not very tall, gets "big pig" vision over the top of the grass on which it feeds, and the most advantageous view of predators approaching across the savanna. So warthogs can and do look without being seen, scanning the horizon at frequent intervals, having their heads up as much as a third of the time.

Scanning is a widely used strategy on the plains. It is highly valued among grazing herbivores, along with a good burst of speed and being part of the largest possible herd. This is known as the "many eyes" strategy—the bigger the group, the better the chance of detecting and escaping from a predator. Other factors, such as burrow availability, provide an upper limit to the size of warthog herds, but the sounders compensate for this by joining forces with other grazers, particularly zebras, who have even better vision and far higher sightlines. When grazing with zebras, warthog scanning rates drop dramatically and their food intake goes up.

The warthog, admittedly, is a special case—a diurnal pig of the open plains. One would expect this species to have better than average eyesight. But to dismiss all the other pigs and peccaries as animals with weak eyes ignores a lot of indirect evidence.

One of the most promi-
nent features of Old World pigs is their
cephalization. Evolution has gone to their
heads. The original form is the relatively
unspecialized Eurasian wild boar, which
has changed very little over 30 million
years. This species has an undifferentiated
head, still in primitive balance with
the rest of the body, with no obvious
difference between the sexes. But in
almost all other species of pigs, males
have acquired much larger heads, which
are clearly pronounced and well set off
from the body. They are also accentuated
by an astonishing variety of growths and structures.

IT IS NOT TRUE THAT PIGS HAVE
poor eyesight. They are
designed to see and be seen.

The point of all these organs is that they make the animal
behind such finery look larger and more ferocious. They are designed to
be seen. They are part of a visual display that is intended to be viewed head on by an opponent
who makes a frontal approach and faces the full effect of such scary equipment. And such confronta-
tions seldom take place in the dark. The performance is carried out in broad daylight in an open
space where the full impact of the appearance can be appreciated by a rival who may, as a result,
even give up and go away on the basis of what he has seen.

Pigs are designed to see and be seen.

WHEN ALL ELSE FAILS AND MERE POMP IS NOT ENOUGH, PIGS OF COURSE DO COME TO BLOWS. THEN
some less obvious armament comes into play. The main weapons in pigs are not extravagant upper
tusks, but the smaller, sharper blades of the lower canine teeth. These are normally covered and are
revealed only when an animal, male or female, opens its mouth. The way it does so and the manner
of its use depend on gross anatomy and have given rise to two very different styles of combat.

The oldest and least specialized is "lateral fighting." It is practiced by pigs with long
and narrow faces, species such as the wild boar and all domestic pigs that have short tusks and no
facial furniture in the form of warts or other outgrowths of the skin. It consists of an initial strut
or jostle in which opponents stand shoulder to shoulder, heads down; or circle each other, bristling
and raising their spinal manes, pausing from time to time to display their broad flanks to the greatest
effect. Then they engage in a trial of strength, leaning on each other's sides, wedging a snout
under a rival's hip, trying for a throw-down. This wrestling can go on for minutes and often comes
to nothing, neither making any attempt to bite, but such disputes can, very suddenly, get serious.

Then both antagonists grind their teeth, snap their jaws, and salivate. And as the
tension rises, each grunts and thrusts at the other with an open mouth, teeth bared as they box, slash,
and side-swipe at each other's necks and shoulders, or bite at an exposed leg or ear. This can result

ALL TRUE PIGS SHARE A SIMILAR repertoire of grunts, squeals, and snorts that express well-being, arousal, rage, fear, and pain.

in deep lacerations and septic wounds and has led to an interesting defense on the part of boars, who engage in fierce fighting during the rutting season that can go on for months. During this time, the boars develop shields of heavy connective tissue across the shoulders and sides that bear the brunt of most attacks. Some are said even to strengthen this armor by rubbing against pine tree trunks to build up a further protective layer of gum.

The second martial art is "frontal fighting," which is the specialty of all the more decorated pigs with thick skulls, bony ridges, and long tusks. It also starts with a warm-up strut, circling stiff-legged, body tense and slightly arched, lowered head pointing at the opponent. This leads eventually to a full frontal approach in which the two adversaries rush at each other, sometimes violently, meeting head on with a crash of skulls, warts, and tusks. These clashes can last for twenty or thirty minutes, with each fighter trying to get past the other's defensive tusks, or to push his opponent backward into a vulnerable position. None of the species that fight this way have shields, so the strategy depends upon prolonging the head-to-head contact as long as possible, wearing the other down until he submits. A defeat is usually signaled by kneeling and squealing, and is acknowledged by the victor simply turning his back and walking away.

This chivalrous, less dangerous, and more structured form of combat seldom leads to serious wounds. Taken together with the organs of display in all the more specialized pigs, it represents a more "progressive" series of adaptations than those of the old-fashioned, less ornate lateral fighters. And this advantage precisely mirrors the situation in all other groups of ungulates. Wild goats with relatively modest horns all fight laterally, slashing with their short horns, just as wild boars do with their 3-inch teeth, and wild cattle, with majestic sweeps of horn meeting in a heavy boss, tend to frontal battles with more formal and ritual strategies, very like the ones used by better-dressed pigs.

The ceremonial threats and charges of the frontal fighters are reminiscent of medieval pageants where knights met in ritual jousts and other well-scripted duels, where honor and chivalry ruled, and being properly turned out was almost as important as being well trained. Long lances and broad swords were as unwieldy and impractical as exaggerated tusks, but that didn't matter as long as the other man was burdened with the same equipment and contained by the same code. The result was a good show and far fewer real casualties.

Peccaries are another story altogether. They have long, narrow faces and short, sharp canine teeth. Peccaries lack any kind of adornment such as warts or shields or bony flanges, and they should, if the pig rules apply, be old-style lateral fighters. But they are not. Fights among peccaries are frequent and do employ some of the lateral strategies. They circle each other, press together, and try to bite one another like lateral fighters; but they also use headlong frontal charges and take advantage of well-developed submissive patterns that avoid most serious wounds.

The answer seems to be that peccaries began as lateral fighters, slashing sideways at each other like archaic pigs with short teeth. But as they migrated into the New World and became more social, traveling and living in larger sounders, and developing wider gapes and more lethal lower tusks, they had to find ways of controlling aggression. They did not go the pig way, becoming sexually dimorphic and creating supermales with ornate hardware and fancy manners. That was impractical for animals who needed to live in large societies with well-balanced sexual ratios. They found another solution—a super-scent organ that produced enough herd-smell to create a cohesive society and sidetrack any individual, male or female, that got close enough to damage another.

You can see it in action. There are lots of little squabbles in big groups of peccaries feeding on scarce resources. These all involve the usual circling, bristling, snapping, and pushing overtures that, more often than not, subside and come to nothing. But if it does go further and evolves into more dangerous in-fighting and whirling around at close quarters, the contestants are automatically brought into such a tangle that they come nose-to-rump together, breathing in each other's heady perfume. Then both stop struggling, their eyes half close, and a soft dreamy look steals over their faces. One sits down on its haunches and nuzzles the other in the sort of submissive posture that gives both of them an honorable way out, without bloodshed.

There's a moral in this somewhere, but probably not one that is likely to do us prideful animals much good.

GIANT PECCARY

Peccaries arrived in South America about 3 million years ago, as part of the "great American Interchange" that allowed ancestral pigs, cats, dogs, deer, and bear to cross the new land bridge at Panama.

They spread out across both continents, becoming at least thirty-four different pig species who battled for precedence against giant ground sloths, armadillos, mammoths, and woolly rhinos. The last of these tall, rangy proto-peccaries became extinct perhaps 10,000 years ago with the most recent ice age. Or so we thought, until 1972.

In that year Ralph Wetzel, a biologist from the University of Connecticut, was making a survey of all the mammals of the Gran Chaco in northwestern Paraguay. This hot, dry, inhospitable thorn forest is still very little known. It is almost impenetrable, consisting of cactus, *palo borracho* (trees so spiny only a fool would climb them), and a hardwood called *quebracho* (the axe-breaker). Rainfall is low and irregular and collects in a few seasonal ponds that look inviting but are full of leeches. The only people who live there are a handful of Indians and a community of German Mennonites, who grow castor beans and cattle.

Wetzel collected a couple of peccary skulls from local hunters, who called them *tagua,* and quickly realized that these differed dramatically from the two species with which he was familiar. They were larger, longer in the face, with enlarged nasal passages and a more rounded snout. And the animals, once he got to see them, were clearly different. They were taller, more rangy, with bigger ears, more delicate legs, and longer, gray-brown body bristles that turned white at the collar.

Back in North America, with access to museum collections of fossil peccaries, Wetzel realized that what he had found was not a new species but something even more interesting, a living population of animals belonging to a species thought to have been extinct since the Pleistocene. He honored the Italian paleontologist who first described the species as a fossil in 1930 by keeping the old name for it—*Catagonus wagneri*—from the Greek *cata,* meaning "downward," and *gonus,* an "angle" of the skull; and in 1975 he published a triumphant paper in the journal *Science,* titled: "An 'Extinct' Peccary, Alive in Paraguay."

It is known now as the Chacoan or giant peccary. There seem to be about 4,000 in the Chaco, where they live in small groups of fewer than six individuals. They

REDISCOVERED IN 1972, the giant peccary was thought to have become extinct more than 10,000 years ago.

are diurnal and browse, feeding largely on cactus pads and flowers, and acacia seed pods. They enjoy wallowing when they can and deal with the hot, dry climate by eating salty soil thrown up around the nests of leaf-cutter ants. Their enlarged nasal chambers help to filter out the dust of the Chaco, and when alarmed they flee through the thorny shrubbery with their long back hairs raised and a spray of secretion from their large rump glands. Pumas have been seen stalking giant peccaries, and it is possible that the milky secretion from the gland, while cementing peccary society, may also be unpleasant to the nose of big cats.

As befits the climate of the Chaco, the skin of this peccary is thinner than that of the other two species and therefore less commercially desirable. But the meat is edible, and the fact that a new, paved trans-Chaco highway is being built is not good news for this endangered species, which also has a smaller brain than either of its living relatives. This is a primitive feature, but it may be offset by the fact that giant peccaries also enjoy the old-fashioned possession of eight, instead of the more modern four, nipples.

There is a captive breeding program in Paraguay, but the best hope for this living fossil lies in the tough nature of its native habitat. It can truly beat extinction only if the Chaco itself is protected.

221270

Hog Heaven

A peasant becomes fond of his pig and is glad to salt away its pork. What is significant, and is so difficult for the urban stranger to understand, is that the two statements in that sentence are connected by an *and* and not by a *but*.

—JOHN BERGER, *About Looking*, 1980

The accepted wisdom about all domestication is that it represents the genetic reorganization of wild animals and plants to suit human interests. Domestication refers to the initial stage of human mastery over the rest of life, our god-given dominion over the beasts of the field, and it is generally considered to have begun in the Old World with people of the late Mesolithic, perhaps as early as 9000 BC. This date, however, is linked to the belief that domestication cannot have taken place until our ancestors had settled down, preferably in some forested valley, near water, and in clearings that could be used for primitive agriculture.

But is that necessarily true? Far from being hungry, nomads frequently had more food than they could eat before it spoiled, and this surplus would certainly have attracted scavengers to the temporary encampments where hides and partly eaten carcasses were left out to rot. It is even likely that such people kept moving because their camps became so popular with other species. But it cannot have been very long before some kind of association grew up between

PIGS ARE EXTREMELY SOCIABLE, enjoying good company—even if it belongs to another species.

these campers and the most persistent camp-followers.

One likely candidate for adoption was a variety of gray wolf, smaller and less wary than its relatives and bold enough to venture into direct contact with our relatives. It would also need to have been persistent enough to stay there and to become   PIGS TASTE A LOT BETTER more docile and tractable by a simple process of selection in which   than dogs do. the more aggressive individuals were chased away or killed, while less toothy, more friendly ones were encouraged or perhaps even captured and kept as pets or as helpers in the hunt. There is evidence of such early, unspecialized canine bones in camp sites in Europe and Asia going back almost 20,000 years, which suggests that "man's best friend" was also perhaps one of his favorite foods.

The full domestication of sheep, goats, cattle, camels, and horses took place much later, perhaps only 8,000 years ago, when our ancestors were properly settled farmers with suffi-cient crops to provide both human food and animal fodder. But there was one other animal that appears to have pre-empted all the breeding and seeding necessary to arrive at such complex levels of Neolithic sophistication, an applicant that came to us as early as the dog and with a great deal more to commend it.

Pigs were ready-made for cohabitation. They almost domesticated themselves, offering us a very generous package. To start with, pig behavior is far more like ours than it is like that of other hoofed animals. Pigs are versatile feeders, good solid omnivores who eat not only what we eat but also what we throw away. Unlike most herbivores, which are almost continuous feeders, pigs eat during selected periods of daily activity and sleep right through the night, just as good humans should.

Pigs are wonderfully accommodating. They enjoy company, even that of other species, and they are not territorial. They are willing to move when we do and to look after their own large litters. They don't need to be herded and are easily trained to come and go, responding to the sound of a horn or a call. And they are smart. Show any pig something just once, and it gets it. Domestication, as far as pigs were concerned, was a good deal—their own idea almost, or at worst a treaty between consenting intelligent parties who entered into the agreement in a spirit of mutual self-interest.

And, of course, pigs taste a lot better than dogs do.

JERICHO, IN THE JORDAN VALLEY, IS THE SITE OF SOME OF THE EARLIEST KNOWN EVIDENCE OF agricultural activity. By 8000 BC, it was a town of 3,000 residents permanently settled inside a massive stone wall. These were true Neolithic farmers with cultivated plots, an irrigation system, and signs of domesticated livestock, prominent among which are the remains of many pigs.

Outside the famous wall, there is even older evidence of flimsy huts built by Meso-lithic hunters who made temporary camps in the area before 10,000 BC, and in their refuse have been found the remains of many butchered pigs with interesting features.

Some of the most frequent finds on archaeological sites are the solid parts of pig skull bones. One of these, the lachrymal bone just in front of the eye socket, has become a valuable

"THE BEST IN THE MARKET."

THE CHINESE KIND.

An unceasing attention to improvement has produced or new-modelled the Chinese breed in this country to what is deemed to be nearer perfection. The delicacy of appearance, the thin transparent ears, small head, short small legs, and even the colour of the hair, are all considered as requisite qualities which ought to be attended to in this kind. They are seldom fed for the same purposes as the larger kind of Swine, being accounted too small for being dried into bacon; but they are preferred as the best and most delicate for pork and roasting pigs.

Our figure was taken from one of this description, in the possession of Geo. Baker, Esq. of Elemore, in the county of Durham.

AFTER EXAMINING A VARIETY OF domestic pigs, Charles Darwin concluded that small, fat breeds originated in China more than 10,000 years ago.

marker because its relative length and shape are a very good guide to the age and size and origin of the owner. Lachrymal bones found at Jericho show that the pigs outside the walls, in the hunters' camp, were larger and more rangy than those inside the walls. Both belonged to the same species, but within just one millennium these pigs had become significantly smaller. That sort of change in such a relatively short time is not natural. It suggests artificial interference of the kind that is characteristic of human selection in the process of domestication.

Similar evidence has been found at the sites of Jarmo and Argissa-Magula in Greece, where other groups of very early settlers appear to have been managing wild pigs long before they corralled their first wild sheep. It was not until 6000 BC, almost into the Bronze Age, that sheep became common domestic livestock.

It is even possible that pigs and people were drawn together long before the Neolithic by a shared passion for something sweet.

One of the mysteries of archaeology is the abundance of grindstones on sites that predate agriculture. There are simple querns and millstones in many early Stone Age encampments that must have been used for grinding grain, but there is no sign of the kilns or ovens that would have been necessary to bake bread. So what did they do with the grain mash? Perhaps what we still do: let it turn into malt, which is richer in sugars and provides the crucial precursor for all sweet, cereal-based drinks, and which, left to ferment, becomes alcoholic. Crushing dry grain is far simpler than grinding it into the fine flour needed for baking, and the main waste product of malt manufacture is a heady swill that usually gets thrown away but would have been very attractive to pigs, something they could smell from miles away.

So before our ancestors cultivated crops, they may well have cultivated a taste for beer, something well worth settling down for, and something that would have provided them with very early and agreeable drinking companions. Dogs aren't the least bit interested in the products of fermentation, but pigs dote on them.

The exact identity of the first pigs to join our party remains a mystery. There are, however, several candidates. In 7000 BC, lake dwellings in the Swiss Alps included lean-to sheds that look remarkably like modern pig pens, built close enough to keep human homes warm in winter; and by 6500 BC, pig remains appear in the domestic garbage of homes at Anau in Russian Turkestan. Charles Darwin, in *The Variation of Animals and Plants under Domestication,* suggested that China was a more likely source of pig paternity, and he could be right. Archaeologists working at Zengpiyan in South China have recently found what seem to be domestic pig bones in a settlement dated to 8000 BC.

SOW OF THE IMPROVED BREED.

By a mixture of the Chinese black Swine with others of the larger British breed, a kind has been produced which possesses many qualities superior to either of the original stocks. They are very prolific, are sooner made fat than the larger kind, upon less provisions, and cut up, when killed, to more useful and convenient portions.

Our figure was taken from a Sow of this kind, in the possession of Arthur Mowbray, Esq. of Sherburn, in the county of Durham. She had a litter of nineteen pigs to support at the time, which was the third within ten months: the whole amounted, in that time, to fifty pigs.

The Chinese or black breed is now very common in England. They are smaller, have shorter legs, and their flesh is whiter and sweeter than the common kind.

A kind similar to this were those found in New Guinea, which proved so seasonable a relief to our circum-

This discovery is still contentious, but there is clear evidence of pig breeding from the late fifth millennium BC in coastal China, and by 3486 BC one Chinese ruler was issuing public decrees encouraging his people to capture and breed pigs and even to use them in trade with other nations.

Tombs of prominent people during the Shan and Zhou dynasties of the second and first millennia BC include many porcelain pigs put in the hands of the dead as possessions worth having in the next world, while in the Han dynasty, around 200 BC, whole pigsties were interred, proof that pig domestication was not only well under way but also an important part of the Chinese economy.

Similarly, in the fertile crescent of Mesopotamia, the site of Tell Asmar near "Ur of the Chaldees," dated to 3000 BC, has been found to be littered with pig bones. The bones consist mainly of those from animals less than a year old, which is a sure sign of domesticity—the spoils of a hunt of wild pig always include bones from individuals of a far wider age span.

The best evidence of pig domestication in the Mediterranean comes from Egypt. In the Fayum depression of the Western Desert, early cultivators planted barley from about 4200 BC, and malted it for their own pleasure, and perhaps for that of their pigs. In all of Middle Egypt during the fourth millennium BC, pigs were not only bred, but also given ceremonial burial. Graves of the period often include glazed figurines of sows with their young.

Pigs also abounded in pre-dynastic Upper Egypt. In 2200 BC, swine were listed among the possessions of an official known as Thutinekht; and in 1980 BC, one Menthuweser was appointed "Overseer of the Swine" to Sesostris I. By the eighteenth dynasty in 1400 BC, even the mayor of El-Kab owned 1,500 pigs; and in the reign of Seti I in 1300 BC, priests in many of the temples of Upper Egypt were breeding pigs on sacred ground.

Pigs have a curious history in Egypt as a whole. In Neolithic times they were exalted, but by 1000 BC there seems to have been increasing prejudice against them. In the Nile Valley, pigs were often depicted in connection with Seth, the long-snouted personification of evil, rival to the sun god Horus, who was blinded by a black pig. When this composite god fell out of favor and his image was effaced from monuments, pig fortunes took a similar dive and swineherds became the most despised class in all Egypt, forbidden even to enter any temple.

This reversal of esteem is a common theme in Egyptian religion. It seems to have been the fate of all sacred animals eventually to become unclean. But while pigs were in favor, they were popular subjects for votive offerings and were widely depicted in paintings and lapidary engravings. In most of these, they are shown as small pigs, rather leggy, with slender heads and pointed snouts, all marks of long-domesticated stock. But a good number are also shown with bristly manes and striped piglets, which are definite signs of the wild.

The best guess about the origin of domestic pigs is one that acknowledges a surprisingly wide range of early experiments by both Mesolithic nomads and Neolithic settlers, in areas around the Mediterranean and as far east as China. It is possible, too, that throughout the warmer areas of Eurasia, such isolated centers traded pigs with one another and drew on a wild stock of just one species that still boasts 16 subspecies scattered across the entire continent.

EURASIAN WILD BOAR

The Eurasian wild boar is the most likely candidate for domestic pig ancestry since it has the largest range of any wild ungulate.

It is classified as *Sus scrofa,* from the Latin *sus,* "a pig," and *scrofa* meaning a "breeding sow." (The word "scrofulous" comes from the same root and means "having swollen glands," usually as a mark of tuberculosis, but referring also to the swollen teats of a sow with young.)

Taking all the subspecies into account, wild boar occupy an impressively wide range of habitats, from tropical rain forest, through temperate woodland, scrub, and grassland, to near-desert conditions—all the way from Morocco to Japan, and from Denmark to Java. They have also thrived as feral introductions to most of the New World and Australia, which pretty well covers the globe.

There is a pattern to this zoogeography. The farther east and the moister the climate, the larger wild boars tend to be, rising to males over 6 feet long and 4 feet high, weighing up to 700 pounds.

Generally speaking, all wild boars are round-backed, long in the leg, with relatively small heads and ears. Their coats are coarse and bristly, brownish and grizzled with white about the face and throat. There are no warts or body decorations, and the upper canine teeth form tusks that curve outward and upward. The lower canines are razor-sharp, honed by rubbing against the upper ones, and the tail is long

with a simple tuft on the end.

Boars everywhere live in maternal family groups, averaging about a dozen individuals of all ages. Adult males are essentially solitary, except when breeding. The boldly striped young are born in litters of four to six after about 110 days of gestation. They spend most of their first two weeks in an elaborate nest or basin of grass, leaves, and branches up to 6 feet long, all constructed by the sow in the hours before she farrows. In less than a month, the litter are following her around

WILD BOAR CAN STILL be found almost everywhere in Eurasia, from tropical rainforest to near-desert conditions.

and trying out solid foods, and by the third month they begin to lose their handsome livery and take on a more conservative dress.

Wild boars eat just about anything: grass, roots, berries, bracken, fruits, nuts, fish, frogs, lizards, snakes, earthworms, eggs, rubbish, carrion, and anything else small enough or slow enough to catch. In the tropics, they have even learned how to open coconuts.

Sounders will forage over a wide area if necessary, traveling up to 100 miles away, but more usually, they linger where food is rich, digging and rooting, sometimes carrying quite large objects, such as marrows or the bodies of young deer, to safer places to consume at their leisure.

Wild boars swim well and wallow with enthusiasm. They are diurnal when unmolested, and everywhere they retire to sheltered resting places some distance from all the signs of their activity in feeding grounds and dunging areas.

Wild boars are lateral fighters, starting with bristling threats and circling, building up in season to parallel-pushing and biting between rival boars. At higher intensities, boars savage each other's shoulder shields or rise up on their hind legs in a vicious "waltz," butting heads and growling and foaming at the mouth. These battles are punctuated by "time-outs," which involve conspicuous yawning and digging movements as displacement activities. Out of season, squabbles in a sounder are settled with short bouts of pushing and shoving, accompanied by a movement called "shaking dead," which seems to derive from snake-catching techniques but transfers very effectively into vivid threat.

During the mating season, males leave salivary gland markers on prominent objects and urinate on the tracks of sexually active females. Courtship involves nose-to-nose and nose-to-genital contact and leads to a rhythmic "love song" from the male that culminates in a direct invitation when he lays his snout suggestively on the sow's rump. The gesture carries with it all the possessive *brio* of a young lover with his arm placed firmly around the waist of his girl in public. Copulation soon follows, and it keeps males so busy in the mating season that they lose up to 20 percent of their weight.

Sounders are normally quite vocal, especially in deep cover where they use at least ten different calls to signify contact, hunger, stress, fear, alarm, warning, food, aggression, and submission. There is a clear hierarchy within a sounder that limits in-fighting. As a result wild boars often live to be twenty years old or more.

BY THE TIME OUR ANCESTORS GOT ROUND TO THINKING ABOUT SETTLING DOWN WITH A GOURD OF grain juice, the Old World was crawling with potential pig partners, but there were problems with most of them.

Warthogs had only four teats, which severely limited the size of their litters. Bush-pigs and forest hogs were not much better, with six teats each. Pigmy hogs were too small, and the warty pigs all lived on remote islands. That left only the wild boars, who seemed to have everything, including twelve nipples and very fast-growing piglets.

So it was not just by default, but because of all its advantages, that almost everyone across Europe and Asia seems to have begun working with wild boars. We may never know where or when it began, but some time between 10,000 and 12,000 years ago groups of our direct ancestors started thinking seriously about getting to know wild boars better.

Sus scrofa was already an all-purpose, unspecialized, ubiquitous, reasonably farmer-friendly wild pig, and because everyone was dealing with the same basic species, all stocks could and obviously did interbreed. Selections from a wide variety of habitats and climates, each the result of particular local human preferences, began to circulate. Slowly to begin with, moving from one local source to another, but accelerating quite quickly as a result of trade and travel, commerce and conquest, descendants of the early stocks became more widely distributed than any other large land mammal.

As a result there are now over a billion domestic *Sus scrofa* in the world, some of them bearing only a distant resemblance to the original wild boars.

Domestication in pigs, wherever it took place, brought about the same morphological and behavioral changes. In the early stages of deliberate breeding, heads became smaller, legs shorter, and bodies more elongated. All these differences reflected the fact that domestic animals no longer needed to fight for supremacy, or run away from predators, or remain light-footed and rangy. They could afford to relax and let themselves go. That is why many modern pigs have long, floppy ears that collapse on to the face and severely impede their hearing. They no longer need to set a premium on being alert, and this feature now gets selected by farmers, because it also tends to make such pigs more docile.

The next stage was an extraordinary change in the shape of the skull. This is evident not only in the convenient lachrymal bone, but everywhere in the face and on the lower jaw. These have been squeezed back in on themselves, providing a profile that is no longer straight or gently curved, but radically broken or "dished," giving all domestic pigs a short muzzle, crowded teeth and a steep forehead.

Such changes don't seem to have had anything to do with the brain. There is no evidence that domestic pigs are any less intelligent than their wild ancestors. On the contrary, their close association with us, and the opportunities we provide, may even have made some breeds brighter in some ways. But the changes in the shape of the brain case and the rest of the skull do directly reflect the fact that domestication means a life free from the need for sexual display or adornment. Battles for dominance disappear along with the need to root or burrow as aggressively as wild pigs. The result is a head pared down to just the basic functions of an undemanding life.

Le Cochon

Le Cochon de Siam

Improved Dorset Pigs.
The property of M.ʳ John Coate of Hammoon, Blandford, and the best pen of Pigs
in any of the Classes at the Smithfield Club Show, 1866.
London, Published by Rogerson & Tuxford, 216, Strand. 1867.

There were changes, too, in the coat. Wild boars have long, coarse bristles and a crest on the spine. Domestication has refined the hair and cut down on its density, to the point where many modern pigs are virtually stripped naked. They no longer need to retain a year-round coat to protect them from the weather. They come instead to carry unusual, nonadaptive, superficial markings that ignore the advantage of being cryptic and can be almost anything at all. And if these new patterns please, they can be fixed in a stock by careful selection. Among the changes that have occurred in domestication is the loss in the young of the bold stripes that once provided perfect camouflage in the wild. Such markings now provide no selective advantage and are seen only as throwbacks in some domestic lines.

BREEDING HAS PRODUCED A RANGE of pigs with smaller heads, shorter legs, longer bodies, and lower manes, but all regress astonishingly quickly to look like wild boar again.

There are, of course, wide variations in the degrees of shortening, broadening, flattening, and tailoring, depending on the domestic stock. But there is one key feature that immediately identifies any given pig as domestic, something no wild pig or peccary ever has—a curly tail. If it is curly, it is man-made, a tell-tail.

The long tails of most wild pigs are expressive—in the warthog, very expressive. They can be twitched and curved or held out away from the body to express arousal or to punctuate sexual and threatening postures. Most wild pigs use them in this way, and some domestic pigs still retain some capacity to wag or stretch their apologetic remnants a little. But tails are now passé, behind the times. They are, in the words of veteran hog-watcher William Hedgepeth, "thin, shabby, disproportionate appendages which seem to be fixed to hogs only because hogs are animals and animals are *supposed* to have tails, no matter how perfunctory."

Or they may just be hedging their bets. Pigs everywhere, left to their own devices, tend to regress. Somewhere in their genes, there exists a store of original wild boar material, a sort of emergency package—"To be opened only in times of dire need"—carrying the basic instructions necessary to become again a free, wild Eurasian animal, complete with a bristly coat, a razorback mane, and a very bad attitude.

This regression is astonishingly quick and can be initiated in a single generation by simulating hardship, by feeding a domestic pig less than it has become used to. Undernourished domestic piglets grow up with longer, straighter, narrower heads than their parents. As adults they look more like their distant ancestors, prepared once more for the hardships involved in finding their own food and fighting for it.

This genetic persistence and plasticity is extraordinary. Domestic dogs that run wild soon revert to pack behavior, but they don't look significantly different. They don't suddenly become more vulpine. But pigs certainly become more porcine. They seem to be transitory domestics, provisional livestock, content to enjoy the benefits of the Neolithic revolution but never quite accepting it as a permanent institution, always keeping their options open. As soon as human largesse stops, it's "Thanks for all the swill!" and they are off to the woods again, ready to root.

You can't help admiring that.

THE FIRST DOMESTIC PIGS WERE ALMOST CERTAINLY KEPT UNDER SEMI-WILD CONDITIONS, MORE OR less free to forage, responsive only to a call at feeding time.

Their situation then must have been very like that of Roman times, when enormous herds of domestic pigs browsed in the valley of the Po or the forests of Tuscany. Sorting out ownership was not a problem. Each resourceful swineherd carried a horn with a tone of its own, and a blast on that set off a stampede that brought his charges, and these alone, trooping out of the fields and woods.

Living this way was economical and ensured a degree of hybrid vigor. The sows were not only allowed to enjoy a little rough trade, they were actively encouraged. They mated readily with wild boars in the forest, dipping into the old gene pool, keeping in touch with their heritage. Each sow could afford to do so because her piglets, no matter how much stamina and hardiness they might inherit from their wayward father, would be as pliant and adaptable as any home-grown litter.

Whatever variation there was in early stocks depended partly on the changes wrought by domestication itself, and partly on a very hit-and-miss process of deliberate breeding based on color, docility, fertility, and the amount of lard a pig could provide. Some of the results were "improvements"; a lot were not. Very often natural selection and artificial selection pulled the pigs concerned in different directions. But things did, in the end, settle down, leaving all domestic pigs in one of two large categories: the European "wild boar" type and the Asian type. The former is agile, alert, and muscular, heavily built, with an arched back and coarse hair, while the latter is characteristically docile, lightly built, with a low back and layers of fat just under the skin.

Natural selection works on a set of characteristics that improve the chances of survival for a species. Human selection, however, works on a more random collection of superficial differences that will not necessarily improve those chances. What results, if it pleases and can be repeated, becomes a recognized and named *breed*.

China alone is said to have produced more than 500 breeds of pigs in its history, but it now houses less than 50. Ian Mason in his *World Dictionary of Livestock Breeds* lists and describes several hundred breeds still extant, and Valerie Porter in her far more accessible *Handbook to the Breeds of the World* illustrates 200 of these modern pigs. No one knows exactly how many breeds of domestic pig actually exist, because there are a large number of local variants whose characteristics are often so evanescent that they leave no trace on the agricultural record and have no real pedigree. And confusion is compounded by the fact that two breeds with the same markings may well have no genetic connection at all. Pigs can't be judged by their coats alone.

Breeds are generally identified by a combination of color, ear carriage, face shape, and general conformation. But the only reliable way truly to identify or trace a pig is through its recorded pedigree, which is kept in a herd book or a breed book held by one of the recognized pig societies that monitor breed features. Unfortunately, these societies were not established anywhere until the late nineteenth century.

In simple terms, breeding involves fixing and enhancing desirable characteristics, while eliminating or reducing undesirable ones. This means juggling with the genes, and knowing something about genetics.

Genes in pigs, and everyone else, come in pairs. They also occur in two active forms or alleles, one of which is dominant, the other recessive. In breeders' shorthand, dominant alleles are indicated in capital letters, such as *D*, while recessive alleles are indicated in the lower case of the same letter, such as *d*. So the gene pair in this example would be indicated as *Dd*, and eggs or sperm produced by such a pig would carry either *D* or *d* alleles, never both. Depending on the inheritance from the father or the mother, there can be four possible combinations of these two alleles in any embryo—*DD, Dd, dD,* or *dd*.

Very few traits are controlled by just a single pair of genes, but ear carriage in pigs happens to be one of these. It determines whether ears will be "pricked," standing up erect, or "lopped," falling over. Lop is dominant and prick is recessive. So *DD* will produce a lop-eared pig—but so will *Dd* and *dD*, because the dominant allele always prevails over the recessive allele. The recessive form of pricked ears can only express itself when no dominant is around, which occurs only in pigs with the combination *dd*. Therefore, three out of every four pigs in such a breeding program are likely to have lop ears. Breeders know this and arrange their matings accordingly.

As far as colors on pig coats go, white is also caused by a single dominant gene. So, for example, the breed known as the large white—a colossal, all-white Yorkshire breed, once described as "too large for any useful purpose"—can be coded WW. Its recessive allele would be ww, which actually manifests itself in a breed known as the large black—an equally huge, all-black pig with oriental ancestors.

But it isn't always this simple.

A cross between a large white pig and a large black pig can sometimes throw up a curious third allele. So Ww or wW may result in a roan pig, with a mixture of dark and light hairs.

This can get even more complicated, but breeders operate on the rule of thumb that white is dominant over black, black is dominant over all spotted coats, spotted is dominant over red, and the gene for having a band of any other color round the body is dominant over not being banded at all. The natural gray-brown of all wild boars, known to breeders as "agouti," is a genetic wild card with all the stubborn qualities of its antecedents. It tends to pop up again on feral animals.

Pig breeders these days have a lot of fun experimenting with unusual crosses, just to see what happens, but there was a time when agriculture was far more conservative. In the nineteenth century in England, pigs in the southern counties were usually black. Those in the Midlands were colored or spotted, and in the north nothing would do but white. This may have been a whimsical preference, but it had survival value. Domestic pigs are very susceptible to sunburn and it makes good practical sense to keep darkly pigmented pigs in the sunny south and pale pigs in the cooler north.

The extraordinary variety of pig breeds is possible because they are more prolific than any other large mammal and reach sexual maturity more quickly than most other species. They are ideal raw material for anyone who wants to raise livestock in shapes and sizes fit to meet the changing demands of fickle markets, and to get them to such outlets in the shortest possible time, in the greatest possible numbers.

This is all wonderful for commerce and trade, but the pig's plasticity is also filled with potential hazards and unexpected hardships.

Just after the First World War, in the tiny village of Piétrain, 25 miles east of Brussels in Brabant province, a handful of farmers managed to produce a strangely spotted pig with hams that bulged in ways that were the envy of body-builders.

As a result of a single mutation, the rumps of these pigs became double-muscled, an excessive enlargement that provided an extremely high proportion of lean meat for fresh pork. These pigs had a lean-to-fat ratio of 9:1, compared to the 6:1 ratio of most other lean breeds. However, in 1920, when postwar customers coveted fat, that was very bad news.

Nevertheless, and virtually unnoticed beyond the bounds of Piétrain, those few farmers continued to breed their muscle-bound pigs, refining the stock into otherwise solid, well-balanced animals with erect ears and black spots surrounded by off-color haloes of gray-white hair. They were not pretty pigs. The hazy, irregular patterns on the coat gave them an undecided look, which only served to accentuate the unlikely, steatopygic bulge of their buttocks. But there is no denying that they were fascinating.

The Piétrain was rediscovered in 1950, and a breed book was opened in 1953. By 1955 it was being exported to France, and in the early 1960s it played a major role in compensating for over-fatty breeds in Germany. The world population of this "Muscle Pig" is now probably about 40,000 breeding sows. It is not endangered any more, but there is a catch. The same gene that provides the extravagant hams of lean meat also saddles the Piétrain with a less desirable characteristic. These pigs are prone, under any kind of stress, to suddenly drop down dead.

It doesn't take much. Being boxed for transport or mixed with strange animals, or even the exercise involved in mating or farrowing, is often enough to make a Piétrain's temperature rise irreversibly. The pig just keeps getting hotter and hotter, until it cooks its own brain.

The condition is known as "porcine stress syndrome," and the gene involved is fortunately recessive. It has been called the halothene gene, because it also responds strangely to the common anesthetic gas of that name. Halothene puts most pigs into a state of complete relaxation, but susceptible Piétrain respond to it with rigid, extended legs. This does at least provide a simple test for the gene, which can be administered to all animals soon after weaning. But the knowledge that any two individuals with Piétrain blood could carry the recessive gene without showing any visible symptoms does put a damper on any large investment in the breed.

Cross-breeding between two animals with substantially different backgrounds can help prevent such problems by making the combination of two recessive alleles highly unlikely. (That is why we have legislation against marriage between people who are too closely related to one another.) Geneticists call this *heterosis* and count on it to provide a certain amount of hybrid vigor. Hybrid sows tend to mature earlier and to produce more piglets, which is a good argument for keeping gene pools fresh and conserving as many different breeds as possible, even if their qualities may be ones that don't seem desirable right now.

In the subtropical basin of the Yangtze River in China there are several breeds known collectively as Taihu. These are caricature pigs, dark-skinned, heavily wrinkled, barely able to see through the folds of fat on their faces, with long, heavy ears like those of bloodhounds. Their growth rates are slow, they have high subcutaneous levels of fats, and they are inclined to be somewhat idle. They flop around like curmudgeonly clubmen expecting to be waited on and they have very few redeeming features—except for the fact that the sows, whose stomachs drag on the ground, routinely produce more than thirty piglets in a litter. The record is forty-two, of which forty were born alive.

A recent crossing between a Shanghai Taihu breed known as Meishan with a large white sow in England produced thirty-seven piglets—twice the usual number for the breed—and thirty-three of these survived to grow into hybrids with increased prolificacy and all the usual gentle social graces of the Yorkshire pigs.

Such a combination of the best qualities of both parents in the offspring is highly desirable. Pig breeders are delighted when genes mingle so well. They call it "nick"—really good nick.

TO BEGIN WITH, THE FIRST DOMESTICATED PIGS LOOKED VERY LIKE THEIR WILD BOAR ANCESTORS, BUT before long, a wide range of regional types developed, partly to satisfy local tastes but also as a result of climate and habitat. In due course, this mix of selective pressures produced a new kind of zoogeography of domestic pigs.

In Northern Europe, breeds tended to be boarish, with long heads and legs, high backs, and big bones. Around the Mediterranean, pigs were smaller, more refined, and darker skinned. Farther east, in Asia, there was a tendency for pigs to be prick-eared, dish-faced, and fat. During the last two centuries, trade and cross-breeding have produced a far more complex global pattern, but there are still regional differences that identify some breeds with their centers of distribution.

In Europe, right from the beginning, there were two distinct races: the ancient Celtic race of the north, with long heads, large tusks, narrow bodies, a crest or "carp's back" of bristles, and long, floppy ears; and an Iberian race in the south, smaller, with long snouts, short tusks, smooth compact bodies, dark or reddish in color, and ears erect, not flopping over the eyes.

These were so characteristic of their countries that they were identified as "land races"—the typical, long, lean bacon pigs of cool northern lands, such as Denmark, and the smaller, fatter, fresh pork pigs of hot southern lands, such as Spain. There was an obvious climatic influence at work here, long before any thought of refrigeration. In both areas, pigs were raised in the open in forest or on pasture, but in cool Scandinavia most of the stock were slaughtered in late autumn and smoked to preserve a supply of meat for the long winter, hence the emphasis on bacon. In warm Iberia, however, sounders ranged freely throughout the year, being slaughtered as needed and eaten as fresh pork before the meat could spoil.

The ancient Celts made good and wide use of their pigs, leaving carvings and statuettes of rangy pigs with floppy ears on standing stones and in holy wells all across Ireland, Wales, and Norway, but it was in Jutland that a long, lean land pig with ears hanging over its eyes gave rise to a dynasty.

In 1879 Magnus Kjoer was managing a bacon factory there, breeding pigs for the lucrative British market opened up by a Scottish steamship line that ran between Copenhagen and Leith. He experimented mainly with large white sows and other imports with the necessary bulk, and he succeeded in producing some useful hybrids, but their yields were disappointing until he looked a little closer to home.

The local Jutish pig is a sorry-looking animal, very long in the leg, straight-sided and lop-eared, with a scraggy dark coat and a hangdog demeanor, the sort of pig you expect to find scavenging on rubbish dumps. But it is as long and lean as a limousine. Kjoer came to it only as a last resort, a pig that was being kept as a curiosity by a few farmers who sold its weaners at rural pig markets. The first crossbreed between a Large White boar and a Jutland sow was, however, promising. So he persevered and kept on trying until he hit the jackpot in 1895 with a boar called Rasmus.

This product of unknown parentage had great nick. Every large white sow he served produced a perfect blend of good lean meat and Jutish hardiness. In just a few years, Rasmus became one of the most famous foundation sires in the pig world. All that was necessary was a little fine-tuning to give the breed better hams, a thicker belly, and less back fat. And there it was, the Danish Landrace. In 1906 it got its own breed book, and it soon became virtually the only breed in Denmark, purveyor to the world of Danish bacon.

Now there are Swedish, Norwegian, French, Swiss, and even South African Landraces, all producing very lean, very long bacon that has to be sold to the high end of the market to compensate for the fact that each animal produces only a limited amount of meat. The Danes are working on that problem too, experimenting again with new and improved Celtic land pigs and taking good care to ensure that on Fyn Island they keep herds of all the old breeds, just in case one of these funny-looking pigs may once again be needed.

The ancient Iberian pigs can make no such claims to commercial success, but they have earned a different sort of notoriety. These red and black, long-snouted lard pigs with their pricked ears became great explorers. They not only formed more than half of Italy's pig population of primitive pasture pigs, but from the fifteenth century onward they went with Portuguese and Spanish explorers to West Africa, the West Indies, and the American mainland. There they spread their genes among the population of pigs kept by the early settlers, and when they ran wild as feral pigs they soon regressed into a formidable New World recreation of the old Eurasian wild boar.

CHRISTOPHER COLUMBUS SAW NO NEW WORLD PIGS ON HIS FIRST VOYAGE, BECAUSE HE NEVER went far enough south to encounter the peccary. So on his second voyage in 1493, he carried not just sheep, goats, cattle, and horses, but also a mini-sounder of eight sturdy Iberian pigs.

These pioneers were taken ashore in Hispaniola and turned loose in low marshland country very much to their liking. As pigs do, they bred, and by 1497 their progeny were running wild through the forests and canebrakes of Jamaica and Cuba as well. Soldiers hunting escaped slaves there found their jungle missions greatly complicated by belligerent boars.

In 1519, Hernando Cortés landed on the mainland of Mexico and burned his boats behind him, thereby committing his entire force to survival through conquest of the Aztecs, but not before he had turned loose an essential herd of pigs. In his 1524 excursion to Honduras, he led a colorful cavalcade into Mexico City on his famous charger Morzillo, followed by musketeers, crossbow men with steel helmets, courtiers in the latest fashions from Madrid and, bringing up the rear, the preferred ration of a drove of Spanish swine.

In 1539 Hernando de Soto set out from Cuba to explore Florida and the east coast of the mainland, carrying "a great many hogs and loaves of cassava bread." On this expedition he was accompanied by 600 soldiers, twenty-four priests, thirteen Spanish sows, and two good boars. By the end of the first stage of his grim march through "sixty leagues of desert" and mosquito-ridden swamps, many of his men and most of the horses had died, and the survivors were reduced to short rations, carrying their supplies on their backs. But their little herd of Iberian pigs lived happily off the land and had increased their numbers to more than 300.

Dining daily on pork, de Soto continued this three-year trek through 3,100 miles of what is now Georgia, the Carolinas, Tennessee, Alabama, Arkansas, and Louisiana, meeting and sometimes fighting the Indians. On the Tombigbee River the expedition survived a battle that cost them eighteen men and 400 swine to a local cacique who had acquired a taste for Spanish pork. Nevertheless, when they reached the Mississippi in 1543 and built a fleet of brigantines to carry them downstream back to Mexico, they still had more than 700 pigs in tow. Most of these they "jerked" and carried on board as provisions. The rest were left with friendly Indians on the river, where they prospered and in time joined the little herds set up all along the route, in camps and in the wild, contributing to a loose association of feral Spanish pigs that still flourish in the brakes, swamps, forests, and badlands of the southern states.

In 1565, Admiral Pedro de Avilés added to this immigration by landing a company of another 400 pigs in southern Florida, a resource that resulted a century later in eight large towns, seventy-two missions, and two royal haciendas, all depending more or less on pigs. Most of these were semi-domesticated, but many strayed or fell into the hands of Indians who preferred to let them roam the woods, where they could be hunted on demand. They still are.

Much the same thing happened in French possessions on the Gulf of Mexico. Pierre le Moyne in 1699 introduced some hardy hogs from Corsica and the Canary Islands. These were pigs not unlike the Iberian breed, but they were jealously guarded and kept apart from the Spanish pigs so that it is still possible to find some of their direct descendants running wild in isolated areas of Mississippi and Louisiana.

British settlers of the New World were a little slow to realize the importance of having pigs as fellow pioneers. Walter Raleigh changed that with his colonizing expedition of 1607, which set up the Virginia Company at Jamestown with three sows that multiplied into a herd of over sixty in their first year. But in 1609, this seed-herd of unknown pedigree was augmented with bold Iberian blood in an astonishing way.

The island of Bermuda was discovered in 1503 and later named by the Spanish in honor of the explorer Juan Bermúdez, but it was given a wide berth for another century, partly because of a dangerous fringing reef but mainly on account of its reputation as a place haunted by spirits whose siren song could be heard on the wind at night. So it remained unexplored and uninhabited until 1609, when Admiral Sir George Somers was forced to beach his flagship *Sea Venture* there after hitting the reef.

All those aboard got safely ashore on a rough coral strand fringed with mangroves and marshes and, instead of sirens, found themselves castaways in the company of thousands of wild pigs—animals left there, presumably, by a previous Spanish wreck. Sir George died not long afterward, but the rest of the company survived, saved from starvation by the pigs. Their escape ten months later in two pinnaces built from the wreck was celebrated in 1611 by Shakespeare, who based *The Tempest* on their tale. He turned Bermuda into Prospero's "Bermoothawes Island" and let his rhymester Rich describe their final escape with the words:

The two and forty weeks being past
They hoist sail and away;
Their shippes with hogges well freighted were,
Their hearts with mickle joy.

In Virginia, the island pigs were added to the British stock on Hog Island in the James River, where they crossbred and flourished to such an extent that their hybrids had to be released, along with other livestock, in the mainland woods. By 1627 it was still possible to count the number of cattle, but the swine were simply "innumerable."

Farther north, on Plymouth Rock and in the colony of Massachusetts Bay, the Puritan Pilgrim fathers were better organized. They too found it difficult to feed ever larger flocks of animals and turned them loose in the surrounding forests, but at Concord they instituted a law requiring Englishmen to mark the ears of their hogs, so that when one was brought to town for sale it could be recognized as belonging to a particular colonist. Any animal brought to market by an Indian was accepted only as long as its ears were unblemished.

Colonists on the mid-Atlantic coast were mostly from Holland, serviced by the Dutch West India Company. Their pigs were of Celtic stock, lean animals probably much like the early Dutch Landraces, shipped to New Amsterdam below decks in pens floored with sand. In America, they were allowed to run loose over most of northern Manhattan Island, but they were separated from the farmlands or *bouweries* and the trading post in the south by a stout palisade. This structure is what gave Wall Street its name, though it is decorated now by images of bulls and bears instead of pigs. But the Bowery remained, at least until very recently, as a rundown area of flophouses and pawnshops where once there was a shady path that led to the farmhouse of Peter Stuyvesant.

By the end of the seventeenth century, there were pig fairs in New York and in Pennsylvania, where William Penn's Quakers were setting up self-supporting farms, but most New World settlers looked only to their own short-term needs. They seldom developed pasture lands or housed their animals or provided winter feed. In autumn, they shot, salted, or smoked a pig or two for their own use. The rest of the stock ran wild, at the mercy of the Indians and wolves. Despite this, pigs flourished everywhere from New Orleans to New Amsterdam (Brisbin 1990). Their new surroundings were much like those of Europe and they took to them gladly, drawing on their reserves of fat and their durable gene pools, reverting rapidly into sounders of razorback hogs with all the old resources of their wild ancestors.

BY THE END OF THE SEVENTEENTH CENTURY IN NORTH AMERICA, FERAL WILD PIGS OCCUPIED ALL the broad savannahs and forests of the eastern states. Later European settlers were offered pork by the local Indians and never wondered from whence it came. All they knew was that the hogs involved were hunted in the wild.

This was the heyday of the free range, a time before the great march west when anything that wasn't fenced was fair game. Times that, in some places, continued until free-ranging was made illegal in the mid-twentieth century. Until then there were great swaths of land with unclaimed populations of feral pigs in at least twenty states. Of perhaps 3 million animals, most were descended from European domestic stocks, but a few were the result of deliberate introductions of Eurasian wild boars during the previous two centuries. Sometimes it is hard to tell the difference.

Florida Rozer Back Hog.

John Mayer and Lehr Brisbin of the University of Georgia have made a painstaking morphological survey of these animals and suggest that they are best described as "wild-living pigs," encompassing feral domestic hogs, introduced wild boar, and hybrids between the two.

IN AMERICA, PURE-BLOODED alien pigs mixed with feral hogs to produce lean, mean, and elusive "razorbacks."

And just to add a little spice to this pig stew, they throw in one further possibility. In recent years, some odd bones have been found among pre-Columbian remains. These all involve true pigs, not peccaries, mixed in with fossil material from other animals that no longer exist. One such specimen from Texas consists of a complete pig skull in which is embedded a large and prehistoric projectile point. Another find from Arkansas has been carbon-dated to 6,000 years ago. There are more problem bones in Tennessee, New Jersey, Maryland, Florida, and South Carolina—all identified not just as Old World pigs but also as domestic variants. The implication is that pigs showing signs of domestic changes were introduced to the New World by some unknown human hand in the late Stone Age, or that the feral animals themselves found their way from an Asian source while a land-bridge still existed over the Bering Strait. Either way, we are pushing back the history of Old World pigs in the New World to the time of the Ice Age, and finding further evidence of the extraordinary persistence of wild pig genes.

It is instructive in this affair to look at one deliberate introduction of Eurasian wild boars to the New World early in the twentieth century. It took place in 1908 in the Great Smoky Mountains, that part of the high Appalachians that lies between North Carolina and eastern

Tennessee. This is a pristine line of peaks more than 6,000 feet high, once the home of the Cherokee Indians and still perceived as the haunt of mountain men who hunt squirrels, make whiskey, feud with one another, and spend most of their time on rickety porches singing sad country songs.

The ownership of the Smokies remains byzantine. Before the existence of national parks, it was vested largely in the hands of truly smoky corporations involved in the timber business. One of these, the Great Smoky Mountain Land and Timber Company, in 1908 sold a huge acreage to the Whiting brothers of England, who made the deal through an entrepreneur called George Gordon Moore. In return for his services, all Moore asked for was 1,600 acres for himself around a bare peak called Hooper Bald.

This unusual arrangement gave rise to rumors of a gold strike in the area, but Moore was a canny man with big long-term plans. He knew a lot of wealthy investors who liked to hunt, and he believed that a luxurious shooting preserve on the site would be even more rewarding than a gold mine. So he built 25 miles of road to his lodge at the peak and enclosed 1,500 acres with a split-rail fence 10 feet high. This extravagance became the wonder of the hollows, and excitement ran high in 1912 when strange, closed railroad cars began to arrive in the little town of Murphy, North Carolina.

Crates from these were transferred to a train of ox teams hitched to wagons that spent days hauling their cargo up to Hooper Bald, where their restless contents were finally turned loose. They were savage "Roosians," furious wild boar with tusks long enough to disembowel an ox. According to one account, "The mountaineers went shinnying up trees like scared squirrels. They were used to wild mountain razorbacks and acorn-splitter hogs all their lives, but these terrible pigs were something altogether different."

In the weeks that followed, the fame of these monsters spread. "They could," it was said, "walk up to a coiled rattler and literally eat him alive, like a morsel." They dealt with the split-rail fence in short order. They destroyed it and came and went through the gaps as they pleased. And the only hunt that was ever organized against them cost the lives of countless dogs and at least one hunter. In 1920 George Moore gave away his empty luxury lodge along with $1,000 and free title to the preserve, which now forms part of the Nantahala National Forest, where the ruins of his dream can still be seen.

The boars, which were said to have come from the Ural Mountains, were later described as of Polish origin. Whatever the case, they invaded the whole of the Great Smoky Mountains in the end, occupying most of the 800 square miles of national parkland and overflowing into Tennessee. These days, the pure-blooded aliens have mixed with feral razorbacks and are now only two-thirds Eurasian, but they are as wild and mean and elusive as ever, and their line and their legend lives on in excess stock trapped and transferred from Appalachia to the Carmel Valley of California in 1923, and to the Aransas National Wildlife Reserve in east Texas in 1931. The US Department of Fish and Wildlife is distinctly dubious about such alien introductions, so guess who picked up the tab? None other than George Gordon Moore who, despite his losses, still seemed to have a sneaking regard for his feisty wild boars.

AT THE BEGINNING OF THE TWENTIETHTH CENTURY, THERE WAS ANOTHER UNRULY POPULATION OF feral pigs in North America. They lived in the Colorado River delta on the border between California and Arizona and they were known as razorbacks.

Razorbacks were described as "fast as horses, shifty as jackrabbits and, when cornered, ferocious as tigers." They did indeed have long legs and enormous, scimitar-shaped, razor-sharp tusks. It was believed they were descendants of Spanish pigs released there in 1886 and that they owed their needle-pointed teeth and suppleness to the fact that they had interbred with wild peccaries. That, at least, was unlikely. But they certainly came to live like javelina, secretively and nocturnally, keeping to themselves in the hard country, growing to look more and more like dark, bristly, wedge-snouted wild boars.

In such circumstances, the changes wrought by domestication melt away rapidly. Sounders reappear, rediscovering the typical group structure of their wild ancestors. They become more mobile, more curious, more wary, showing pronounced tendencies both to fight and to flee. Their flight distance and level of aggression are greater, particularly where rival males are concerned. In females, the varied and complex patterns of nest building are soon revived, and piglets are weaned far sooner and learn to fend for themselves more quickly. Sexual behavior becomes more specific too, responding only to the proper stimuli, which is in marked contrast to boars and sows who can, in desperate domestication, be aroused by simple dummies or hormone sprays.

The archetypal razorbacks of the South are as much the product of their environment as they are of their genes, and their pedigree, although Spanish, is unknown. But there is at least one population of feral pigs in the New World whose heritage is clear. They live on the barrier island of Ossabaw, 15 miles from Savannah, Georgia, and they are the direct descendants of pigs brought from Spain in the 15th century—typical Iberian medieval animals, with prick ears, heavy coats, and long snouts, bred to scavenge and survive on their own once the Spanish had gone. On the island, they have had no contact with any other pigs for almost four centuries, and they have never had to worry about predators. For 200 years, the island owners have protected them and their habitat from change, and the result is fascinating.

For a start, they are the smallest feral pigs in the world, standing not much more than 18 inches high and weighing on average about 60 pounds. This is not a surprise, because there is a tendency for island populations of all mammals to be small. Dwarf elephants once lived on islands in the Mediterranean, and Shetland off the coast of Scotland is renowned for its tiny ponies. The Ossabaw pigs are similarly rather short-legged, perhaps because they have no need to flee for their lives. They tend to be somewhat aggressive among themselves, a consequence perhaps of being overcrowded. The optimal population on the island is around 500 and, from time to time, the owners do thin out the numbers: 900 were trapped and removed to the mainland in 1990.

No attempt is made to feed the Ossabaw pigs. They are left to fend for themselves and, as a result, they have retained most of their other ancestral characteristics. They are typical dark-pelted, sharp-faced, foraging pigs bearing unusually long bristles with split ends. But their most interesting adaptation to recurrent periods of near famine on sandy Ossabaw is an ability to build up back fat in good seasons, laying down a reservoir almost 4 inches thick. This makes them, proportionately, the fattest of all wild-living mammals in the world. These pigs also, however, have unique fat-metabolizing enzymes that release energy to keep them going when times are hard. Associated with this useful ability is a low-grade, non-insulin-dependent form of diabetes that is now the subject of research by a group studying human obesity at the University of Georgia.

Professor Dennis Sikes at that institution says: "Man is more nearly like the pig than the pig cares to admit." Maybe, but it seems in some respects that pigs are far better at adapting to the environment than we now are. Being short and fat is valid adjustment to life as they find it.

The best measure of the importance of not having to worry about predators comes not from an island but from a whole continent without indigenous carnivores. In 1788 Captain Arthur Phillip arrived in Australia with a few pigs on board his vessel as fresh meat. He turned the survivors loose in New South Wales, and by 1799 there were 3,459 of their offspring in the colony. Just six years later there were too many pigs to count, and Australia was well on its way to a feral pig problem.

The immigrant pigs scattered widely from the south, up the east coast to Queensland and on into the Northern Territory, dealing with problems as they met them and developing under the influence of natural selection into new varieties. The pigs on Kangaroo Island are descended from a single pair left there in 1801 by the French navigator Nicholas Boudon. White with splotches of dense black hair, long snouts, and strong feet, they are, following the insular pattern, very small, and they have tiny litters. To help deal with the vicissitudes of island life, they have also acquired the kind of vigor that results from having a higher red blood cell count than any other pig in the world.

New Zealand has its own dwarf pig. It is called *kunekune*, which means "fatty" in Maori, and it has a delightful nature. It is docile, slow-moving, easy to please, and fattens on nothing but grass. These chubby little charmers once wandered freely through Maori villages, following their owners about, always up for a nudge or a tickle. They came, it seems, through Polynesia, growing smaller with each move away from the Asian mainland, and they have all the characteristics of old oriental stock—short faces, dished heads, snubby snouts, and fleshy wattles under the chin. Their fat, mixed with honey, makes a fine tropical substitute for butter.

When you think about it, being welcome in human homes is another very canny porcine adaptation with high survival value.

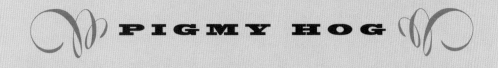

Dwarfism is not unique to islands, at least not to islands surrounded by water. It can also occur in areas isolated just by their geography.

The smallest pig in the world, a quarter the size of the *kunekune,* occurs only in very restricted areas of the Himalayan foothills of Asia. The pigmy or pygmy hog was first described in 1847 by the British zoologist B. H. Hodgson, who created a new genus for it—*Porcula,* from the Latin *porculus* for a "porker." His choice of the species *salvanius* is also of Latin origin but has been the source of some confusion. It could refer to the site of the type specimen in the Saul Forest of Nepal, or to the Latin root *silva* for "forest." But the most likely derivation appears to have been a misspelling of the name Silvanius, the Roman god of the woods.

The pigmy hog is undeniably tiny. Less than 16 inches long and 10 inches tall at the shoulder, with an average weight of 10 pounds, it is about the size of a wild boar piglet. But there the resemblance ends. In shape and color it looks like a large eggplant, dark and glossy with a plump rounded bottom that tapers forward to a narrow, stalky snout. The hindlegs are slightly longer than the forelimbs, giving the impression of an animal poised for instant nervous flight, like an artillery shell on spindly legs.

Its general appearance is closer to the streamlined New World peccary than the sturdy Old World pig,

prompting Hodgson to describe it as an intermediate between Sus and Tayassu. Others suggested that it might even be an ancestor of some of the dwarf breeds of modern East Asian "miniature" pigs, but this misapprehension has now been traced back to the fact that one of the first captive pigmy hogs displayed by the Berlin Zoo, and photographed and featured by Bernhard Grzimek in his 1969 *Animal Life Encyclopaedia,* was in fact not a pigmy hog at all, but the piglet of a dwarf domestic breed similar to the Vietnamese pot-bellied pig.

THE PIGMY HOG IS THE smallest pig in the world, just 10 inches tall and weighing an average of less than 10 pounds.

This confusion has now been cleared up. The genus Porcula is invalid, and the pigmy hog has been reclassified as *Sus sylvanicus*. It is considered to be the seventh recognized species of the genus *Sus* and, despite its external appearance, it shares every internal feature of its dentition, cranial anatomy, and morphology with the other six Old World pigs in the genus. The only major ways in which it differs are in having just six teats, and in being at the dwarf extreme of pig sizes.

The pigmy hog has never been domesticated. Its original distribution seems once to have stretched from Nepal in the west, through Sikkim and Bengal, to Bhutan, but it now appears to be limited to just two small areas of northwestern Assam. Its natural habitat is the "thatch-lands," small pockets of tall-grass savannah or *terai,* which are irrigated by the early monsoon rains. These limited fertile patches are now being rapidly converted to paddy cultivation and are subject to communal dry-season burns, hounding the little pigs into ever-decreasing, discontinuous thickets of secondary forest, where they may number no more than a few hundred survivors.

Pigmy hog behavior is typically piggish. They live in small, mixed family groups of females and their young, numbering anything from six to nineteen individuals. Their gestation period is thought to be about 100 days, and the litter size ranges from two to six. They are very tactile animals, rooting sociably together. Such sounders are not territorial, but they build the most complex nests of all pigs, stacking broken thatch into oval mounds around well-lined chambers where the animals pile up on top of one another to keep warm through the alpine nights. Studies in captivity show that these little pigs sleep very well.

The sexes differ very little. Boars tend to be a little more solitary in the rut, and they fight with rival males only when broadside threats, raised bristles, curled lips, and teeth-chomping fail. Then they throw up clods of earth and charge frontally at each other, but being somewhat ill-equipped for aggression or defense, their encounters never seem to come to much. Pigmy hogs would rather settle for a bout of mutual grooming, which is solicited by lying down, or even rolling on their backs with their feet in the air.

Despite their delicate, pointed snouts, pigmy hogs root actively, turning over litter and topsoil. They eat roots, tubers, grass, fruit, and seeds and seem very fond of grubs and earthworms. They also consume unexpectedly large quantities of earth, which may account for half their total intake. No one knows why, but it may have something to do with the fact that they carry a heavy load of ectoparasites in their thick coats. Among the parasites is a unique sucking louse called *Haematopinus oliveri,* after the British zoologist William Oliver, who has made the only extensive study of pigmy hogs in captivity and in the wild. This infestation may also explain why these little hogs are so passionate about being groomed.

Pigmy hogs are critically endangered, both by poaching and by loss of habitat. They were written off altogether in the 1950s, and they barely exist now in the sanctuaries of Manas and Barnardi in Indian Assam. It would be a great shame to lose them and their genetic resource. We still have much to learn from them about pig evolution, and exactly why mainland Eurasia should have produced just two living pig species (or three if you count a recent claim to have rediscovered the extinct *Sus bucculentus* in Vietnam). At one extreme there is the hefty and aggressive wild boar with at least 16 subspecies covering every other habitat on the continent; at the other, a tiny, reclusive, rodentlike relative, clinging to the edge of the Himalayas, just longing to be groomed.

THE EURASIAN WILD BOAR STANDS LIKE A COLOSSUS OVER THE HISTORY OF DOMESTIC PIGS. NO ONE doubts its stature as an ancestor or its presence in every pedigree. The genetics are straightforward. The same nineteen pairs of chromosomes can be traced in an unbroken line from the wild stock to nearly all domestic breeds, though it is interesting to note that, in yet another demonstration of pig plasticity, domesticity may also have made genetic changes that reduce the chromosome number in a few breeds to eighteen pairs.

Through all the years in which domestic breeds have been proliferating, their progenitor has been there, on its parallel track, holding its own, just biding its time in case it should be needed yet again. And now, it seems, the time has come. In several countries, wild boars are being "farmed," by selective hunting, by the creation of special reserves, or simply by tolerating their existence on the fringes of our lives. We have begun again to treat the wild boar as if it were part of our livestock system.

Wild boar meat is low in fat and cholesterol, a gourmet dish with a distinctive texture and flavor. In France and Italy, close to a thousand farms have been set up to harvest and even breed wild boars commercially. The prodigal swine has returned.

We can't take all the credit for this strange reversal. Boars are very difficult to fence in or out, but in the last thirty years they seem to have reintroduced themselves to areas in which they were once plentiful, without benefit of feasibility studies, expensive radio-tracking devices, or any of the tedious committee meetings that normally precede such ecological programs.

This event, unplanned and unexpected, was abetted by some unusually severe weather in the 1980s that produced changes in the countryside and rearranged wild boar patterns of movement, but the happy result has been an unintentional self-fulfilling improvement in biodiversity and agricultural economics, and the creation of a new and welcome wildlife display. It is, in short, a conservationist's dream, the only downside so far being just three cases in which wild boars have forced their way into farm units to mate with domestic sows.

This must have been how it all began over 10,000 years ago.

There is very little information about the first nine millennia of pig-keeping, but it is possible to get a glimpse of how things stood between pigs and people in the last millennium. The best records of this are those that deal with the history of the pig in Great Britain.

To begin with, 2,000 years ago, "pig" meant an animal not unlike the wild boar—a long-legged, razorbacked, dark brown, and bristly beast with bad manners. This hybrid was a little smaller perhaps, but still a rangy animal, only slightly less difficult to deal with than its formidable ancestor. It retained upright pricked ears, and nobody who owned one dreamed of trying to change it in any way.

The Iron Age pig, by all accounts, had ten or twelve teats and probably produced an average of eight piglets in each litter, which is three more than a wild boar. That would have been very welcome, but any other change would have been unwanted and impractical in a system in which most pigs were killed and salted down in early winter. Even the best farrowing sows were, it seems, slaughtered after their third litter. A reliable boar might have lived a little longer, but most domestic pigs never got to celebrate their second birthday.

The medieval pig was mainly a poor man's animal, providing him with the only red meat he was likely to eat, and with pig fat, which was vitally important in an age when extracting oils from vegetable sources was difficult or impossible. Pigs could easily be reared on poor land or no land at all, and in the country the animals could be turned loose in the forest to find food for themselves without competing with their owners for limited resources. Weaners also provided revenue and a trade opportunity for people who had no other source of income.

This situation, however, was too good to last. From the twelfth century on, the wealthy began to impose restrictions on pig-keeping. Forests were cleared to provide grazing for landowners' sheep, access to pastures and commonage was denied, and the woods that remained were taxed.

Pigs belonging to the poor could only be fattened by an autumn feed on acorns, beech nuts, and other forms of wild mast, so every forest was classified according to its pig-carrying capacity, and "pannage"—the right to allow pigs to forage—was imposed by feudal lords and church estates. This was paid for in kind, usually a proportion of the pigs slaughtered in the autumn, or in cash. By the reign of Henry VIII, pannage was set at the rate of one penny per pig, and any failure to pay was severely punished.

The institution of such practices had begun with the Norman conquest in the eleventh century and was facilitated by the *Domesday Book*, William the Conqueror's extraordinary survey, or "Description of All England" in 1086, which identified every land, landowner, and occupant, down to the humblest tenant. This made it possible to tax people and their pigs and changed the way in which almost everything worked.

With the introduction of a tax on pigs came the first professional pig husbandry men in Europe, the village swineherds. These were men who collected groups of pigs from the various households every morning and tended them during the day. Their reward was a sucking pig each year and the entrails of any animal slaughtered, but there is no mention of any attempt by these overseers to do anything more than walk the pigs to and fro and protect them as far as possible from predators or pig poachers. They were, however, legally responsible for all animals in their care, and there are many records from the twelfth century onward of fines imposed for damage done by loose pigs to crops and property.

The herds involved were sometimes enormous, but none of the owners kept more than a few pigs and there were no signs, even up to the sixteenth century, of anyone setting up or running a large commercial venture with pigs. The nearest thing to it would have been someone who owned a good boar and rented him out to owners of promising sows in return for a small fee, or for the favor of first choice from the resulting litter. The only hint of any awareness of breeding for desirable qualities comes in records from which we learn that, wherever there was a choice being exercised, it was white pigs that were preferred.

Up until the thirteenth century, pigs were largely poor country people's hedge against hard winters and a welcome source of occasional revenue. But that changed when populations grew, forests declined, and pigs became expensive in the sense that they were no longer semi-wild, eating almost for free in the woods instead of competing with humans for available resources.

The result was the "cottage pig"—an animal, often single, permanently housed in a pen or yard, and fed almost entirely on kitchen waste.

COTTAGE PIGS WERE PART OF A DOMESTIC, RATHER THAN A COMMERCIAL, ECONOMY, SO WE HEAR little of them until the seventeenth century, and even then, they tend to be linked in some way to other economic activities. Pig-keeping was advised, for instance, for those whose "other occupations furnish a plentiful supply of food at a trifling expense; as the keepers of brewers, millers and so on. The very refuse of whose customary produce will serve to keep a considerable number of these useful animals." Pigs converted disposable malt grain into fresh flesh that had a good market, but their own refuse was a nuisance in the cities and in London was often consigned to the bottom of the Thames. Townspeople nevertheless housed pigs in their backyards.

Friedrich Engels was startled to find in 1845 that there were flourishing piggeries in every back street of Manchester. "In almost every interior court such pens may be found, where the inhabitants throw all refuse and offal, on which the swine grow fat; and the atmosphere, confined on all four sides, is utterly corrupted by putrefying animal and vegetable substances." He was particularly scandalized to discover that most of these pigs belonged to Irishmen "who love their pigs as the Arab his horse, with the difference that he sells it when it is fat enough to kill. Otherwise he eats and sleeps with it, his children play with it, ride upon it, roll in the dirt with it, as any one may see a thousand times repeated in all the great towns of England."

It was certainly so in London. In North Kensington in the mid-nineteenth century, pigs outnumbered people three to one: "They had their sties mixed up with the dwelling houses. In some cases pigs have been found even inside the houses, and under the beds." In South London, houses were declared "unfit for the keeping of swine," with the result that "pigs were wont to haunt our streets and roads in search of food of the most loathsome and disgusting kind." The town pigs clearly had to go, but in the country cottage pigs persisted well into the middle of the twentieth century.

IN THE EARLY SEVENTEENTH CENTURY, MOST PIGS IN BRITAIN LOOKED ALIKE. THEY WERE CALLED "Old English Pigs," large, gaunt, bristly brutes with the profile of broken-nosed bruisers. They were coarse pigs, dark as the devil, with many of the worst traits of the wild boar, and the only way in which they obviously differed from the medieval pigs was in their size. In town and country, they ate more, exercised less, and got fat. A drawing of one by Rembrandt shows a solid, long-bodied animal with a very shaggy, unkempt coat, but by his time, pigs were already showing some signs of "improvement."

The first of these changes was to their ears. Whereas Celtic medieval pigs had the erect pricked ears of the wild boar, Old English pigs had floppy, hanging ears, the mark of the Landraces of Europe and probably introduced to England by the Danes. Pigs with long ears tend to mature earlier and to lay down more fat, two very desirable characteristics.

By the eighteenth century, pigs no longer ran loose; their breeding could be controlled. As a result, there was increasing interest in cultivating finer, fatter, faster-growing pigs,

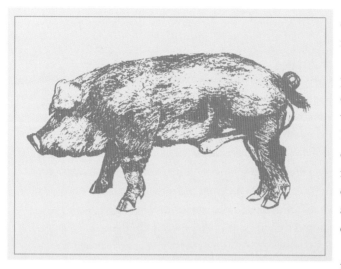

even if that meant importing foreign stock.

Oriental blood had probably already been introduced 1,500 years earlier with the Roman occupation of Britain. Now, in the eighteenth century, it was renewed with the import of Chinese breeds that offered finer form, higher fertility, a nicer disposition, and an extraordinary range of colors.

The only conclusion that can be drawn from Victorian pigs is that they bore very little resemblance to their eighteenth-century ancestors. Applied science and happy accidents had changed them out of all recognition. Where once all British pigs were black or brown, now they came in a bewildering number of colors and patterns, few of which could be relied upon as identifying any particular breed. And it was precisely this confusion that led at the end of the nineteenth century to the establishment of herd books and breed books kept by societies whose main task was, and still is, to fix and enforce the characteristics of each registered breed.

IN DRY, WARM CLIMATES, BREEDERS still favored tough, black pigs such as the Berkshire and the Duroc.

If there is any definite trend in this pattern of development through the eighteenth and nineteenth centuries, it is one in favor of whiteness. This was perhaps inevitable, given that the new white was the exact opposite of old wild boar black and that white is the least practical of skin hues for pigs living out of doors. Only rich people could afford the luxury of white pigs that needed to be pampered, and such exclusivity has always been attractive to men of means. Their selection was largely artificial, but it did have the lucky advantage of a pair of genetic benefits. The gene that makes pigs white just happens to be linked to the production of good white meat, and white in pigs is dominant over all other colors.

Breeders in the north of England were particularly fond of white pigs. Their ideal was big, soft, clean pigs that looked like giant marshmallows with upturned noses and a short leg at each corner. These huge pigs, sometimes actually called Yorkshires, were good in every respect and have been an enormous success all over the world, especially in countries with extensive bacon industries such as Denmark, Sweden, and Poland. Large whites have also formed the basis for countless other breeds abroad, notably the Landrace pigs, and in a smaller version known as middle white, they have done very well in Japan, where there is even a memorial and a shrine erected to one particularly prolific boar. All in all, the large white in all its incarnations is to the global pig industry what those very familiar black-and-white Friesans from Holland have become to the dairy cow business.

Today most of the idiosyncratic regional breeds, many produced purely for the amusement of a single owner, have disappeared. Varieties came and went along with breeds that were never more than a local conceit. In the twentieth century, pigs became commodities in an industry that is still more interested in productivity than it is in character. Now, Britain, like most other modern mass producers, concentrates on hybrids drawn mainly from stocks of large white, large black, British Landrace, Berkshire, Tamworth, and middle white pigs.

A POPULAR TREND IN THE eighteenth and nineteenth centuries was for big, soft, white pigs such as the Yorkshire.

In the rest of Europe, there has been a similar emphasis on good white meat. The large white and the Danish Landrace represent about two-thirds of the producing and breeding pigs on the continent. After the Second World War, Sweden took up the slack left by Germany, and during the 1960s everyone catered to the new fashion for lean meat. Almost the only exceptions were extremes such as the ultra-lean and muscular Belgian Piétrain and the super-fat, sausage-making Mangalitsa from Hungary. There was little enthusiasm or demand for local breeds, though the mole-headed Limousin in France, the red Iberian in Spain, and the white German Edelschwein continue to be popular with gourmets and delicatessens.

China has been the exception. For several millennia the Chinese have been producing peaceful local pigs that concentrate their energies on laying down fat instead of wasting it on foraging.

Most Chinese prefer pork to any other meat, and they prove it by supporting half a billion pigs, more than half of the world's pig population. Some of these are now hybrids with large white and Mandrake that provide a little lean meat, but most of China's pigs are still small, fat breeds that flourish in 200 million peasant households. With the exception of a few hardy, high plateau pigs that have obvious wild boar blood, all Chinese breeds are unmistakable. They are small, short, roly-poly, sway-backed pigs with deeply dished profiles and snub noses. Most have straight tails and all have bellies that drag along the ground.

The rest of pig history, however, is set in the Americas.

OLD WORLD PIGS MADE THE MOST OF THE FREEDOM THAT THE NEW WORLD HAD TO OFFER AND TOOK to the woods, where they multiplied with enthusiasm. But they did even better under management.

The colonies at Worcester, Boston, and Salem in the late eighteenth century had far more pork than they needed, and they barreled their surplus and began to ship it back to

Europe. Virginia alone sent 60,000 barrels to Cuba, Madeira, and Portugal in 1774. Then came the Revolution.

The War of Independence lasted until 1783 and, as usual, the pigs were the first to recover. Pig production went straight back into business, but in a new way. Before the war, settlers had fattened their pigs largely on mast foraged by free-ranging animals. After the war, with new markets in their sights, they began to force the pace and supplement pig food not just with homestead waste but also with surplus Indian corn. In 1790 British pork in Jamaica cost the equivalent of 14 cents a pound. The Americans sold their pork at half that price, with just as much profit, and moved into industrial mode. In 1792 alone, they exported 38,500 barrels of pork at $2 apiece. The profits from this trade were considerable, and some pioneers used their earnings to join the great tide of emigration that was fueled by land laws that promised every family a half-section of good western land. With them went their indispensable swine.

The pigs of the time were well suited to traveling. They were not the pampered pigs of Europe, but tough, adaptable migrants accustomed to living off the land. They were long in the leg, short-bodied, and slab-sided, with rough hair, capable of defending themselves against predators and more than ready to play their part in the winning of the West. They were called "stump-rooters," "snake-eaters," or "wound-makers," among other things, but they always came when they were called, and they trotted along behind the wagons in all weathers. Piglets, as they arrived, traveled for their first weeks in a wooden crate slung from the back axle, next to the crate that held the chickens.

The farmers followed trails blazed by backwoodsmen along the Ohio and Shenandoah valleys, by way of Cumberland Gap or the Great Kanawa. They skirted the Appalachians and the Alleghenies and were ferried across the Muskingum and Wabash rivers by flatboats and keelboats, always following the setting sun, keeping going until they found what they were looking for. In the process they added Ohio, Indiana, Illinois, Michigan, Missouri, Iowa, and Wisconsin to the list of states, and they celebrated every successful stage with a feast of ham and eggs.

The pigs themselves feasted almost every day. Their travels took them through country not unlike the temperate woods and pastures in which their ancestors had evolved back in Europe. So they dined on strawberries, blueberries, hazel and hickory nuts, acorns and roots, cane and reed, wild rye and white clover. But what pleased them most was Indian corn wherever it could be found. It was this that fattened them best and turned these pioneer pigs and their descendants into "corn on the hoof."

For many of the early homesteaders, the West was closer than they expected. By the time they got to the Mississippi Basin, they found the kind of country that offered them not just room to breathe but also land fertile enough to feed their livestock and themselves. It was the runaway success of their hogs that stopped them from dreaming about ranches in Oklahoma or gold in California. As a result, early in the nineteenth century, a strange and unexpected migration took place. Surplus pigs began to be driven back toward the settlements on the seaboard by farmers or by professional drovers who marched the herds to market in Charleston, Virginia,

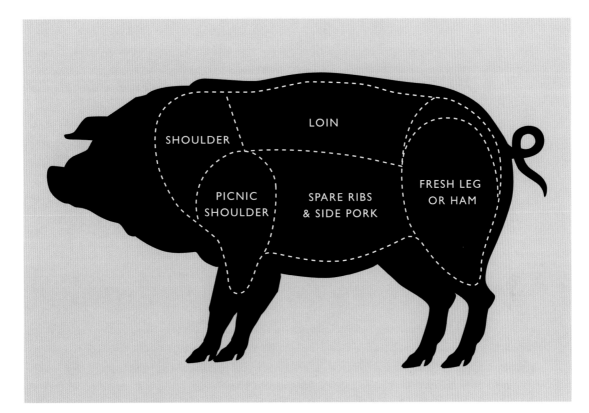

or Boston. These long-distance drives of slab-sided pigs carved out well-worn hog trails as clearly marked and as famous as the cattle trails of the Southwest. They ran by way of Dunkirk to New York, from the Ohio River to Philadelphia, and from Cumberland to Baltimore, all routes that later became the tracks of the New York Central, Pennsylvania, and Baltimore railroads.

A CHOICE PIG NEEDS TO BE ABOUT eighteen months old, weighing in at around 200 pounds, which yields just 60 pounds of fresh pork cuts.

In their heyday, between 1800 and 1850, the long-distance pig drovers prodded something like half a million animals a year back east, where they were swiftly processed into barrels for shipment to soldiers on both sides of the Atlantic, who were provisioned very largely by salt pork. In the War of 1812, so much pork was supplied to American soldiers by a New York packer called "Uncle Sam" Wilson that he came to personify the entire government and was portrayed by cartoonists as a giant figure in a tall hat under a banner that read "Uncle Sam is feeding the Army."

The pig drives were run by men who, on their own, regularly walked 40 miles in a day. Behind a herd of a hundred pigs or more, they were limited to about 10 miles each day, prodding laggards with a hickory staff and keeping up the pace with the crack of a long snake whip. There were "drove-stands" at regular intervals on all the main trails, places where up to a thousand pigs could be corralled and fed overnight, while the drover got a hot meal and a place to spread his blanket in front of the bar-room fire. The pay for those who helped on the drive was about $12 a month, plus half a dollar each day for food and lodging on the long walk home.

It was a hard life, and a risky one, with rivers to swim, ice to break in the winter, torrential storms, thieves and toll-keepers to outwit, and awkward animals whose value at the other end was in the hands of speculators who met the herd a few days short of the market. But when famous trails like "Old Pike," which ran through the Cumberland Gap, were threatened by railroads in the second half of the century, there was strenuous opposition to them from drovers, stand-keepers, and farmers who all insisted that it was cheaper, safer, and more convenient to transport pigs on their own four feet. In the end, the dispute was resolved by the decision not to send live pigs east by either route but to slaughter and prepare them in the west, right where the corn grew.

In the days before refrigeration, every farmer knew how to butcher his own pigs. It was an old art and an event practiced by an owner and his friends, usually in the cool months, most often out of doors.

The chosen pig would be about eighteen months old, weighing around 200 pounds, which yielded about 60 pounds of fresh cuts and another 60 pounds for curing. Killing the pig required some preparation, access to several useful tools, and a willingness to deal with a noisy and bloody execution.

Pig-sticking—the oldest technique for killing a pig—is still the best and the most productive method. The pig is roped, rolled on its back and held a little off the ground by a loop around one hind leg. The aim is to sever a major blood vessel with a sharp knife without damaging any meat or muscle. This requires a steady hand and knowledge of the exact position of the carotid artery, which lies diagonally across the neck three inches below the skin, right beneath the point of the breast bone. A firm, neat cut driven down the midline to the backbone does the job, producing a gush of heavy, pulsing blood that can be collected in a large pot or bucket. Blood-letting takes several minutes, and stirring constantly with a wooden spoon prevents clotting and leaves a bonus of enough raw material for many black puddings.

This clinical description makes it sound easy. It is not. William Cobbett in 1828 wrote: "To kill a hog nicely is so much of a profession, that it is better to pay a shilling for having it done, than to stab and hack and tear the carcass about." The noise and struggle involved in killing a pig are heart-wrenching, but the presence of a specialist, who can handle the pig with respect and stun and dispatch it with surgical skill, does bring an essential dignity to the ancient drama.

The next stage is inevitably messy and is known as "singeing" or "scalding." At its most rudimentary, it involves stacking hay around the carcass, setting this alight and brushing off the burnt bristles with a stiff broom. A tidier solution is to lower the pig from a pulley into almost boiling water and dunk it until bristles begin to come away on their own. Then the task can be finished with knives and bell-scrapers that give the entire pig a very close shave.

The rest is butchery, once described as the "Art and Mystery of Butchering" and best left to the professionals who package "everything but the squeal." In the country or the colonies, they were recognized "fleshers" or just someone with a one-horse cart who hawked his meat from door to door, wrapping his wares in that wonderful coarse brown "meat paper" that was universal before the plastic age.

In the American West, there were butcher-farmers who fattened and slaughtered their own pigs. In the East, there were city butchers who bought pigs alive from the drovers and sold the meat in a public market. But the first large-scale industrial slaughterhouses for pigs were established in the Midwest, on the "Corn Belt."

Indian corn or maize produced a revolution in pork production. From the moment American pigs were first fed on corn, everything changed. They loved it, and it wasn't long before farmers noticed how well they did on it, losing some of their lean scraggy looks and plumping up nicely in all the right places. And as soon as there were surpluses of corn, pig numbers also began to soar. The corn that couldn't be sold was turned into something even more desirable—lard.

Lard comes from the subcutaneous adipose tissue or "back fat," which is deposited mostly on the rump as pigs grow, and in the nineteenth century, it was indispensable. It illuminated log cabins, softened leathers, and lubricated appliances. Mixed with petroleum it produced grease, and combined with lye it made soap. It protected meat from spoilage, preserved a whole range of foods by sealing them from air in wide-mouthed jars, and catered to everyone's health and pleasure until twentieth-century dietitians spoiled the fun.

All animal fats and oils tend to be flavored by their source. So animals fed on onions or fish smell like onions or fish. Pigs fattened on forage from the woods have a gamy flavor, but corn-fed pigs have fat that has a delicious sweet taste. And pork lard, in comparison to all other animal fats, contains more oleic acid, making it easier to melt and tastier for baking. That is why the Midwest is blanketed with limitless acres of golden corn.

This development started slowly. There was a ready market for pioneer pork, but the razorbacks were not very efficient at fattening and didn't provide nearly enough lard to satisfy contemporary cravings for it. Breeders therefore turned their attention to the creation of a national herd with more commercial properties.

The British were already improving their native pigs, cross-breeding with smaller, fatter Chinese pigs such as the Meishan, which Sanders Spencer in 1910 described as "bladders of animated lard." But while English breeders were trying to change their big old, lop-eared pigs into medium-sized, somewhat fatter animals, what America needed was a lard pig that also retained as much size as possible.

The first and most obvious solution was to import some of the ready-made European breeds—pigs such as the black Hampshire, a hardy old English favorite, with its white belt and good reputation as a lard pig. Later came the big white Yorkshire pigs, already well on their way to becoming international favorites as reliable pork providers, and the Danish Landrace. All these are still popular in the United States as modified imports.

More imaginatively, however, new local hybrids were designed specifically for the American market. The first was a rangy old black pioneer pig improved by the Shakers in Ohio and originally known as the "Warren County Hog." Renamed in the 1860s by a Polish farmer who called it the Poland China, this is a superb, meaty pig with white stockings that was once shown at a record weight of 2,552 pounds.

A second hybrid occurred along the Delaware River, where English white pigs were cross-bred with Chinese pigs. This was later refined in Pennsylvania into a large, fertile animal with huge floppy ears that was christened the Chester white.

The most interesting innovation, however, was a ginger-colored charmer with Iberian ancestors that was improved by a breeder in New York. Henry Kelsey owned a thoroughbred stallion called "General Duroc" and when his prize pig grew almost large enough to challenge the horse, he decided to call the new pig breed Duroc as well.

So, by the second half of the nineteenth century, the slaughterhouses of the Midwest had sties full of large local pigs, all dependable lard fatteners and breeders, ready and able to turn the Corn Belt into the Hog Belt—a porcine melting-pot that became the home of a great new commodity known as "pork bellies."

THE ROLE OF THE AMERICAN PIG WAS TO MARKET AMERICAN CORN. THE HOG BELT WAS NOT INTERESTED in selling choice cuts or sausages to a gourmet market. It was instead the greatest corn-and-hog growing industry on earth, involving 4 billion bushels of the nation's gigantic annual crop, and the mass producer of a single commercial product, a side of good lardy meat that came off the line in quantities great enough to be quoted on the stock market. At its peak, the Hog Belt involved more than 3 million pig breeders, producing $3 billion worth of meat—more than was realized from the sale of all other livestock combined. It was an enterprise so large that canals were dug and rail tracks were laid specifically to connect the slaughterhouses with the Great Lakes and the Eastern seaboard. Before refrigeration, slaughter was seasonal, but with the introduction of cooking, smoking, and other forms of processing on an industrial scale, it became a year-round business. Most of the industry was centered on Cincinnati, which became briefly known as "Porkopolis" and was famous for "packing fifteen bushels of corn into a hog, packing the hog into a barrel, and packing a barrel into a train or a flat boat." In 1861 Cincinnati was killing 400,000 hogs a year. Then, during the Civil War, nearly all industrial pig activity was transferred to the relative safety of Chicago, the new Porkopolis, where it remains today.

Agricultural economists draw up interesting charts that identify the centers of activity for each crop every year, and these show that in the United States, hog centers have generally been about 150 miles west of human population centers at any one time. That pattern holds, even today, with the center of swine husbandry in Iowa, where 14 million pigs are being raised almost exactly 150 miles upwind of the markets of Chicago, probably because the pig farmers at each turn were encouraged to take their herds as far as possible from urban centers. This meant going deeper and deeper into the undeveloped West.

After China, there are more pigs in the United States than in any other country, and the United States exports twice as much pork as China, most of it to Japan in the form of processed meat. The US export trade in pork is currently worth $2 billion a year, which is big business, however you look at it.

It all began with the Homestead Act of 1862, which was based on the philosophy that each head of a family was entitled to a home on a farm of 160 acres in return for contracting

to live on and work that land for at least five years. Within a decade, millions of immigrants from Germany, Scandinavia, Scotland, and Ireland settled in the Midwest, occupying 300,000 square miles of fertile land. They were vigorous people, thrifty and quick to learn. Living up to the requirements of the contract, paying off starter loans, and making ends meet was not easy, and the only ready solution was to plant corn and keep pigs. This they did, and by 1880 the pig population in the area had risen to 50 million animals.

GIVEN THEIR ANATOMICAL AND physiological similarity to humans, it was no surprise when pigs were cloned in 1971 for medical research.

At that time, a good sow could be bought for $5. With proper care she could produce three or four breeding gilts a year that would farrow in their turn the next season. In the third year these joint litters could be worth as much as $30 at market, and the settler would still have the original sow. It was the only reasonably safe investment in homesteading, and hogs soon came to be known as "mortgage lifters," the saviors of many an owner. Very often, "hog money" was all that kept the trading store and the local bank off a homesteader's back. With pigs, a farmer could afford to keep a few dairy cattle, and at harvest time, all he had to do was turn the pigs out into standing corn. It was called "hogging down," a wonderful labor-saving device in which the pigs not only fed themselves but also trod down and rooted up all that was left, leaving a field that was fertilized and plowed, ready for planting the following season.

Not everyone prospered, but the combined produce of thousands of small farms in Nebraska, Minnesota, Iowa, Missouri, Indiana, and Illinois kept the pig production lines

in Chicago going through the nineteenth and twentieth centuries, turning what used to be the poor man's meat into a major money-earner involving 3 million farmers. The value of pigs in the United States now amounts to over $5 billion, helping to supply a world that consumes 75 million tons of pork and 2 billion pounds of lard each year, 20 million portions of which are supplied in cans of Spam.

THE MODERN CRAZE FOR PET PIGS like this began in the 1920s with jazz singer Josephine Baker, who performed with a troupe of perfumed pigs.

These days everything is automated. Pigs are assembled on pig farms and disassembled in a packing plant. There they get sorted out into 18 percent ham, 16 percent bacon, 15 percent loin, 12 percent fat back, 10 percent lard, and 3 percent each of spare rib, plate, jowl, foot, and trimmings. In addition, the stomach wall is used for sausage casings, pepsin is extracted from the peritoneum, and mucin is taken from the intestine. The pituitary gland provides medication for arthritis and hormonal problems. The kidneys make a good stew, knuckles pickle well as bar-room snacks, and the brains are sought-after delicacies. And everything listed above goes into pork sausages, which are so popular that the annual global production in links would easily stretch from here to the moon and back, three times.

THE VAST MAJORITY OF THE 1 BILLION DOMESTIC PIGS ON THE PLANET GET EATEN, BUT THERE ARE A modest number whose destinies are very different.

There have been, ever since Babylonian times, a few carefully selected and trained sows whose sole task it is to track down subterranean truffle fungi attached to the roots of broad-leafed trees. The Greeks and the Romans used them, and "truffle pigs" are still in service in France, where they help to harvest the "black diamonds" of Périgord. There have been some recent, scurrilous reports of German efforts to develop an electronic truffle detector, but the suid section of La Fédération Française des Trufficulteurs has threatened to bring the entire rural workforce out in protest if there is any attempt to bring such abominations anywhere near *La Grande Mystique*.

Some pigs postpone conversion to pork by taking up another trade. A black sow in the New Forest of Hampshire showed so much promise that a game-keeper trained her to become a "hunting pig," who pointed at partridge in the coverts as assiduously as any dog. She was named "Slut" because of a tendency to wallow in bogs, but she never forgot her training and she stood to rabbit forty strides away, her nose in an exact line, motionless until the game moved. Her achievements were celebrated in the *Sporting Magazine* in 1842.

Other opportunities in trade have been explored by pigs that have succeeded as "seeing-eye hogs," far more visible than dogs and much less likely to drop everything in favor of chasing cats; "sheep pigs" with more authority than most dogs can muster; "walking pigs" that can cultivate a high-kicking strut every bit as elegant as any show horse; and "rodeo hogs," who just love to entertain. But a pig's best chances of promotion are probably in the army. An agency called Animal Behavior Enterprises has undertaken a study for the US government and reported that "military pigs" could carry surprising loads and are better than most draft animals at living off the land. They could also be easily trained in tracking and mine detection.

The stumbling block to most pig employment in such careers is one of discipline. Pigs are not very good at taking orders. They are too intelligent for that, and they don't take kindly to boring, repetitive tasks. They are rather like humans in this respect, and this has led to a new career opening. Thousands of pigs are now employed in biomedical research.

PIGS ARE RELIABLE BABYSITTERS and natural entertainers, said to have a love of country music

Such studies began in the sixteenth century with Philippus Aureolus Theophrastus Bombastus von Hohenheim, better known as Paracelsus—"better than Celsus"—(Celsus being a renowned first-century encyclopedist). The later version was a young, prematurely bald, obese and tipsy physician in Renaissance Vienna, who nevertheless pioneered modern medicine with a long list of innovations. Among other things, he scandalized his contemporaries by saying that "a pig's composition is very similar to that of man."

The idea was echoed four centuries later in 1953 by Christian Barnard, then a resident surgeon at Groot Schuur Hospital in Cape Town, who remarked, "Strange as it might seem, in several anatomical aspects the pig is closer to the human being than any other animal." A few years later, he joined the University of Minnesota in Minneapolis as senior cardiothoracic surgeon and noted that "even the ape's heart is not big enough to provide a human with sufficient circulation, but the pig's is. It may become the salvation of mankind." He was instrumental in helping the university's Hormel Institute to set up a program of specialized breeding with the aim of developing a "mini pig" that would weigh about 140 pounds fully grown, approximately the size of a human. By the early 1960s, they had such an animal and, as a direct result of experiments with its heart, Barnard in 1967 performed the world's first human transplant back at his home hospital in South Africa.

Pigs' hearts and arteries follow the same patterns as ours. Their bodies require the same kind of food and process it in the same way, and they suffer from the same peptic ulcers under stress. We are both omnivores and tend to grow fat and sedentary, and we share a fondness for alcoholic beverages. Pigs' teeth are deceptively like ours and make perfect models for dental studies. Pigs get cancer, rheumatism, and arthritis as we do, and they respond to drugs and radiation treatment in much the same way. Pig livers have saved the lives of dozens of humans by "porcine perfusion," in which pig and human livers are linked for several hours. Grafts of pig pieces, such as mitral heart valves, are becoming common in human patients.

The list of things we have in common goes on. There are similarities in our gut enzymes, our endocrine system, our immune system, kidney architecture and function, pulmonary structure, respiratory rates, tidal volumes, and many aspects of our social behavior. Pigs are ideal medical models, substitutes for us in research on the structure and function of eyes, the physiology of the skin, and the development of the brain. There is hardly any medical discipline that is not already benefiting from studies on specially bred pigs that can cram a generation into a single year, providing answers to the use of new drugs and procedures that would take 20 years to assess in human subjects.

As a result, pigs are on the cutting edge of transplant technology. In 1992, Astrid, the first genetically modified pig, was born with organs redesigned to make them even more

Ils pullulent les petits cochons, il y en a partout.

humanlike. And in 2001, the world's first cloned pigs arrived with promises of an even better chance of avoiding the immune response that usually blights xenotransplants with violent rejection. We still cannot graft whole pig kidneys or livers to human patients, but in 2002 a Mexican team succeeded in inoculating a seventeen-year-old diabetic girl with cells from a newborn mini-pig that have allowed her to go without insulin or any other drugs for a year.

As if interchangeable organs were not enough, a new line of research in Massachusetts has opened up the possibility of taking immature animal cells and injecting these into the brains of human patients suffering from wasting diseases such as Parkinson's and Huntington's chorea to take the place of diseased cells. These immigrant and restorative cells, of course, have come from our alter egos—from pigs. Pig-cell surgery is bringing us closer together in ways that make it almost impossible to tell who among us is all human or at least part pig.

The tide rolls on, as does a growing discomfort about the ethics of breeding animals for human spare parts, even for the 60,000 patients waiting for transplants in the United States alone. It does seem a little hard that pigs, whose plasticity has allowed them to evolve all the necessary adaptations for meeting the demands of natural selection over the last 40 million years, should be penalized for that very ability. But there is a bright side to the story.

During the same time that miniature pigs have been modified for medical research, a separate stream of the little hybrids has taken the world by storm. The assault began in North Vietnam in the 1960s with the export of several "I" (the breed's name), the native pot-bellied pigs, to Western zoos. These were smallish pigs, maturing at about 150 pounds, but by the 1980s they were being bred down to about 40 pounds, standing less than 14 inches tall.

The New World stock of pot-bellied pigs can be traced back to just 18 little pigs imported from Sweden by Keith Connell in 1985. These have multiplied and have been reinforced and refined by additional new stocks each year, and there are now something like 50,000 short-nosed, round-bellied, black or piebald house pigs trotting about suburbia on leashes or riding around on the handlebars of bikes wearing baseball caps. George Clooney has one. The "yuppy puppies," some of which have changed hands at $20,000, have arrived and are being serviced by several pet pig magazines and a journal, two rival registers, and a Potbelly Association.

There is no telling where this might end. Pigs are very accommodating. A century ago, G. K. Chesterton exercised himself on the future possibilities of pigs as pets:

A monstrous pig as big as a pony may perambulate the streets like a St Bernard without attracting attention. An elegant and unnaturally attenuated pig may have all the appearance of a greyhound. There may be little, frisky, fighting

pigs like Irish or Scots terriers; there may be
little, pathetic pigs like King Charles spaniels.
Artificial breeding might reproduce the awful
original pig, tusks and all, the terror of the forests—
something bigger, more mysterious, and more
bloody than the bloodhound. Those interested in
hairdressing might amuse themselves by
arranging bristles like those of a poodle. Those
fascinated by the Celtic mystery of the Western
Highlands might see if they could train the bristles
to be a veil or a curtain for the eye, like those
of a Skye terrier, that sensible and invisible Celtic
spirit. With elaborate training one might have
a sheep-pig instead of a sheep-dog, a lap-pig
instead of a lap dog.

Most of this has already come to pass, and much more is highly likely. I look forward
with interest, and some cautious trepidation, to 2007, the next Chinese Year of the Pig.

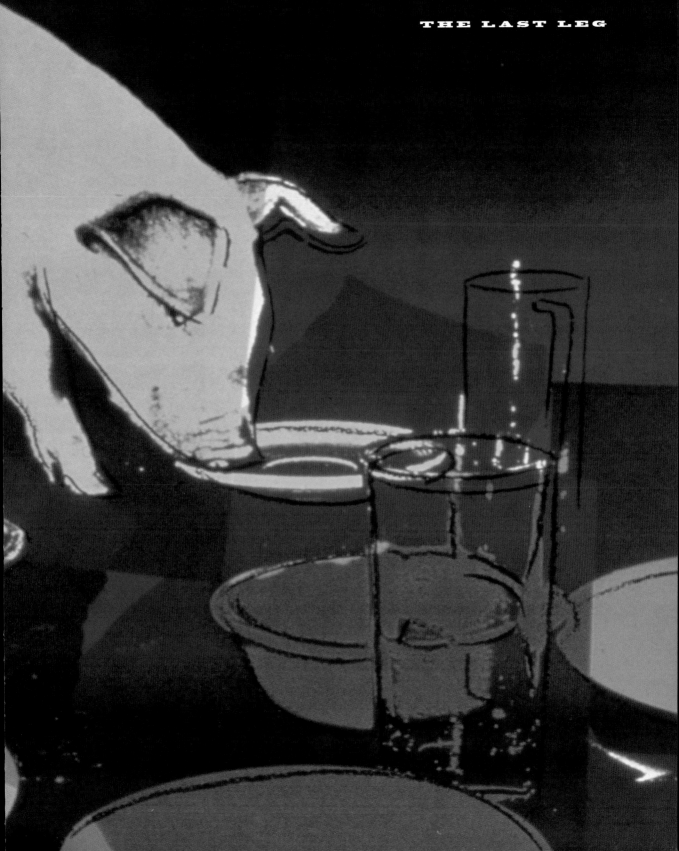

Hogging the Limelight

I have a friendly feeling toward pigs generally and consider them the most intelligent of beasts, not excepting the elephant and the anthropoid ape. I also like his disposition and attitude toward all other creatures, especially man. He takes it for granted, or grunted, that we understand his language, and without servility or insolence he has a natural, pleasant *camarados*— a hail-fellow-well-met air with us.

W. H. HUDSON WAS A GREAT naturalist, but he was not averse to a good rasher of bacon with breakfast.

—W. H. HUDSON, *A Traveler in Little Things*, 1921

Hudson, as always, occupies the high ground, unafraid of being a little more effusive than less sensitive observers. And, as usual, he sees things and feels things that most naturalists miss, or are unwilling to admit.

I have no such qualms. I have been watching pigs for more than half a century, under domestication and in the wild, and I am left with disturbing images of something unfinished, half seen and half repressed, a surprise waiting to be sprung. There is an undeniable feeling of kinship, of something profoundly democratic going on behind the pig's eyes that, though hard to define, is impossible to dismiss.

Orwell's pigs failed, because they were "only human" after all. But real pigs are not so easily set aside. There is something cryptic about them, a mystery waiting to be resolved, a sense of intellectual potential that will not be denied, no matter how hard some people try to relegate them to the farmyard as ignorant "oinkers."

Pigs are aware beings, individuals with minds of their own, waiting, it seems, for us to make the next move, to pose the proper question. But we keep on missing the point, unable to get around the confusion that lies between "pig" and "pork," between being conscious and being bacon.

Louis Bromfield, novelist and professional farmer, was acutely aware of the dilemma, warning: "Look at pigs over a fence but never bring one into your life, for when you put an end to his existence you'll forever after suffer from memories as cannibal and murderer." It is very hard to dine with and dine upon a friend, regardless of the number of his legs.

Concern about this impasse is not new. The Greek biographer Plutarch in the first century AD wrote an essay entitled "Do Animals Reason?" It takes the form of a classic dialogue between Odysseus and one of his crew who has been changed into a pig by the sorceress Circe. The pig attacks the practice of eating flesh and, simultaneously, the Stoic argument denying reason to animals. "The nature of beasts is chiefly to be discerned as it is, neither void of reason or understanding. For I do not believe there is such difference between beast and beast in point of intelligence and memory as there is between man and man." He convinces Odysseus of the moral superiority of many animals over humans and, in the end, decides that he would rather stay as a pig.

Classical Latin poetry included animal epitaphs, involving dirges to gnats and Monty Pythonesque dead parrots. These were largely tongue-in-cheek laments, parodies of the self-important orations of the time, but they were still doing the rounds in medieval times in the form of "Milesian Tales." The *Testament of a Piglet* is perhaps the best known. It consists of the reading of his will by a piglet known as "Grunnius Corocotta"—the grunter—who makes bequests of all his various body parts to people in selected professions. It is very broadly ribald, relying on wordplay and suggestion, alluding to pigs and people, porcine noises and pork products in ways that result in "guffaws amongst curly-headed school boys." And it works still at that level as an entertainment that disguises our unease about killing and eating a named animal.

The English essayist Charles Lamb met the subject head on in his *Dissertation upon Roast Pig* in 1824. Writing at a time when pig breeding was all the rage, he absolves us all of blame by setting up a Chinese scapegoat called Bo-boy, who is fond of playing with fire and burns the family home down around a litter of pigs. In dismay, he tries to rescue at least one piglet and burns

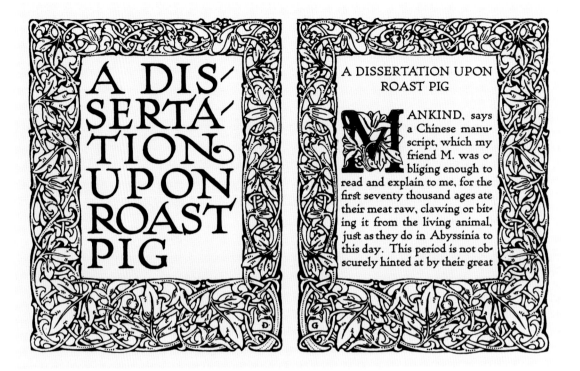

A DISSERTATION UPON ROAST PIG

MANKIND, says a Chinese manuscript, which my friend M. was obliging enough to read and explain to me, for the first seventy thousand ages ate their meat raw, clawing or biting it from the living animal, just as they do in Abyssinia to this day. This period is not obscurely hinted at by their great

his fingers. "To cool them, he applied them in his booby fashion to his mouth. Some of the crumbs of the scorched skin had come away with his fingers, and for the first time in his life (in the world's life, indeed, for before him no man had known it) he tasted—crackling!"

ESSAYIST CHARLES LAMB allowed people everywhere to enjoy gilt without guilt.

Bo-boy's discovery leads not only to an epidemic of burning houses across the length and breadth of China, but also to a catharsis that allows people everywhere to enjoy pork without guilt: "Crisp, tawny, well-watched, not over-roasted, crackling—the pleasure of overcoming the coy, brittle resistance—indefinable sweetness—the tender blossoming of fat—cream and quintessence of the child-pig's yet pure food, a kind of animal manna of fat and lean, so blended and running into each other, that both together make but one ambrosian result." It is, in short, a real indulgence—absolution through living well.

Which brings us to P. G. Wodehouse, Lord Emsworth, and the Empress of Blandings who has her own bijou residence in the grounds of the castle. The Empress is a Berkshire sow and winner of three consecutive medals in the Fat Pigs class at the annual Shropshire Agricultural Show. In the course of several tales set at Blandings, this paragon of animals is stolen, pignapped, hidden, and even saved from dieting. She looms large in the life of the ninth earl of Emsworth, like a balloon with a curly tail, and serves not just as a narrative device but also as a measure of the caliber and quality of all new arrivals. Being "unsound on pigs" is regarded as a fundamental flaw in character, tantamount to a capital crime. Lord Emsworth's greatest delight, the sweetest of all music to him, is the sound of the Empress restoring her tissues, wading through the 57,000 calories required each day to keep her in the full bloom of health. He dotes on her but is at the same time perfectly content to enjoy bacon with his breakfast.

There is no sign in the 1930s of any qualms about all this. That had to wait until Orwell in 1945 and Roald Dahl in 1959. Dahl's short story, simply entitled "Pig," carries no social message and features no pigs but is nevertheless a very troubling critique of the slaughterhouse. The hero is a human being called Lexington who is raised in the country by an eccentric vegetarian aunt and who goes to New York with all the innocence of an animal going to slaughter. In the city, he tastes pork for the first time and, in one of Dahl's surprise twists, ends up being dragged to his death on the same conveyor belt that shocks, squeezes, stuns, sticks, shaves, and slices pigs on their way to market. Dahl is not arguing against meat eating or necessarily campaigning against pig farming. What he does do, however, is to "animalize" Lexington and highlight the thoughtless, dispassionate process of the pig disassembly line.

IN CONTRAST, THE NATURALIST WILLIAM HENRY HUDSON IN 1919 WROTE OF THE DEATH OF A particular pig in a village on the Wiltshire downs, a pig he befriended just days before its death, despite the fact that he too liked pigs "in the form of rashers on the breakfast table." The pig he came to know was kept in a small, filthy sty, "belly deep in fetid mire," so he took to stopping there each day for a talk and a vigorous scratch with his stick, which "made the pig wriggle his body and wink and blink and smile delightedly all over his face."

Wondering what else he could do to make the animal's life less miserable, Hudson picked a bunch of elderberries in a field nearby and offered them to the pig, who looked at the unfamiliar fruit doubtfully but eventually ate it all, and on the following day he hailed the naturalist enthusiastically. From that day on they shared the strange fruit together. "It was a new sensation in his life and made him very happy, and was almost as good as a day of liberty in the fields and meadows, and on the open green downs." On the day of his slaughter, the pig was carried off, leaving Hudson to reflect: "In a month or two, if several persons discovered a peculiar and fascinating flavour in their morning rasher, it would be due to the elderberries I had supplied to my friend the pig, which had gladdened his heart for a week or two before receiving his quietus."

Hudson was a great naturalist, as he proved in his evocation of the countryside in *Far Away and Long Ago,* about growing up in the Argentinian pampas, and in *A Shepherd's Life* and *A Hind in Richmond Park,* both set in England. His love of natural detail and "little things" did a great deal to foster a back-to-nature movement in the early twentieth century, but his most influential work was *Green Mansions,* a romance set in the Amazon, featuring a mysterious forest spirit, half bird and half human. Like his contemporaries Kafka and Joyce, he used animals to reflect an image of modern humankind.

John Steinbeck was another fine amateur naturalist. *The Sea of Cortez* reflects a biologist's outlook on life, but it also reveals a darker vision. By one means or another, all of Steinbeck's animals are put to death in graphic detail. Pigs and chickens are slaughtered, dogs and rabbits struck by moving cars, and turtles executed with an axe. In *To a God Unknown,* he tackles nature worship and rites of fertility through an encounter with a wild boar that "sat on its haunches and tearingly ate the hind quarters of a still squealing little pig." And in the *Grapes of Wrath,* the Joad brothers slaughter pigs quickly and efficiently: "Tom struck twice with the blunt head of the ax; and Noah, leaning over the felled pigs, found the great artery with his carving knife and released the pulsing streams of blood."

This is more than literary realism. Steinbeck describes the death of animals with the detachment of an anatomist, and this allows him to deal with human suffering in passages of unusual tenderness. The simple folk in his stories survive by killing other animals when they have to, and they do so in ways that not only accomplish a necessary goal but also earn respect.

Steinbeck's heirs include William Golding, whose castaways in *Lord of the Flies* encounter evil in the form of the "Beast," realized in the head of a pig jammed on to the end of a pointed stake. "The head remained there, dim-eyed, grinning faintly, blood blackening between the teeth. The Lord of the Flies hung on his stick and grinned. At last Simon gave up and looked back, saw the white teeth and dim eyes, the blood—and his gaze was held by that ancient, inescapable recognition." The head is the perfect symbol in a parable about reversion to savagery.

Pigs are very good at framing our metaphors. Our kinship with them lets them stalk our dreams, haunt our memories, and invade our fantasies in an alarming number of ways. Bruce Chatwin in *On the Black Hill* allows the twin sons of a Welsh farmer to adopt a runt piglet and christen it Hoggage, but their father has no use for runts and one day leaves Hoggage's carcass hanging from a hook in the meal shed. "Both boys held their tears back until bedtime; and then they soaked their pillows through." They never forgave their father for the murder.

In *A Day No Pigs Would Die,* Robert Newton Peck deals beautifully with the same dilemma, one that faces farmers every year. This is a haunting coming-of-age story set in Vermont about the relationship between a young boy and Pinky, his piglet, from spring to autumn, when he must listen to "the strong crushing noise that you only hear when an iron stunner bashes in a pig's skull."

The biological presence or absence of pigs is just one way in which they can capture our imagination. Another is as a trope or metaphor, an animal of the mind. These are not essential beasts, but socially constructed creatures that have become part of human culture—hence "dirty pigs," "greedy hogs," and "rotten swine." In this currency, even a "perfect pig" is seen to have all the characteristics of the animal that are understood to be disagreeable.

The third way of experiencing pigs is as psychological animals, somewhere between animals in body and animals in mind, the kind of beings that are most at home in our dreams and are more easily captured by poets than by zoo keepers. Pigs of this nature are totemic animals, the familiars of shamans and the stuff of Jungian archetypes. They are the Black Pigs of Set, the Calydonian Boar, and the Gadarene Swine, the monsters of our "animalizing" imagination.

Pigs like this are to be found in the boar cults of the Celts, which identify fearsome warriors as boars with giant tusks, and fertile women as sacred sows. Demeter, the Greek goddess

Varâhavatâram

S. वराहावतारं

WHEN A DEMON DRAGGED EARTH to the bottom of the ocean, Vishnu took the form of a boar in order to rescue our planet in battle.

of crops, was identified with pigs. She, and other earthly deities, were depicted wearing pig masks or figurines, and such images have been found at sites where sacrifices were made to the underworld. In the Pacific, mythologies have been built around the sacred triangle of pig-woman-moon at the heart of fertility rites and ceremonies.

There is an extensive global literature on "sacred pigs," much of which links them to menstruation and the magical powers of women, which is a probable source for many sanctions. The way in which the flesh of the pig is so roundly condemned and tabooed in Leviticus is a sure sign that it was once a deity. And the fact that human flesh is reported to taste very much like pork is more than enough to forbid that, too, quite independently of any natural or historical constraints against cannibalism.

Muslims have a jaundiced view of all pigs, but Hindus celebrate Varaha, the great primal boar, as the third incarnation of Vishnu, bearing in his four hands the attributes of a mace, a conch, a lotus blossom, and a *chakra*. Several Christian saints also seem to have had a soft spot for pigs. Saint Kevin rescued a wild boar from the hounds of Glendalough in Ireland, Saint Blasius consorted on several occasions with wild pigs near his mountain refuge, and Saint Paul Aurelian tamed a wild sow in Brittany. The fourth-century Saint Anthony seems to have been most frequently associated with swine. He is usually pictured with a pig at his feet, a well-kept, plump pig fattened by his monks to the point that a proverbial expression of disgust among hungry parishioners was to say the local priest was "as fat as Saint Anthony's pigs."

Saint Anthony is nevertheless the patron saint of swineherds, and his image also occurs alongside pigs with flaming brands in their mouths, signifying "Saint Anthony's Fire"—a bacterial inflammation of the skin called erysipelas. The only logical connection between pigs, streptococcal skin diseases, and a holy man who sought solitude in the Egyptian desert is that the saint is said to have healed such topical infections, which do sometimes respond to applications of pig fat.

Some ecclesiastical pigs get to church in their own right—as relations of the wild boar in Psalm 80, feeding on the Tree of Knowledge and rooting it out of the ground, or propping up misericords, "acts of mercy" in the form of ledges that allowed the elderly to prop themselves up, half seated, during long services through which the rest of the congregation were expected to stand.

Sows and their litters also appear on decorative bosses or on moldings over arches, in honor of the white sow that led Saint Brannock to the spot in Devon where he was able to build his monastery, or the pigs who led Baldred to the discovery of the spa at Bath. Virgil tells how Aeneas, the son of Aphrodite, was advised to build his new city in Italy on a site where he would find a white sow with 30 piglets. These helpful pigs were presumably incarnations of the sow-goddess Phaea, "the shining one," who features in both pagan and Christian legend.

But just as often, pigs found their way onto sacred ground as warnings against lust and greed. In most bestiaries, pigs are said to signify "sinners and unclean persons or heretics; penitents who have become slack and have an eye for those sins which they have wept for; unclean and wanton men or spirits; foul thoughts and fleshly lusts from which proceed unproductive works; and the sinner of good understanding living in luxury."

THE MOST COMPASSIONATE AND EVOCATIVE WRITING ABOUT PIGS HAS ALWAYS BEEN IN LITERATURE intended for children. The earliest, mostly anonymous, pig tales, however, are not ones with happy endings. Little Betty Pringle's pig, "when he was alive he lived in clover, but now he's dead and that's all over." Tom the Piper's Son was equally unlucky: "The pig was eat, and Tom was beat." Revisionary do-gooders insist that this particular animal was only a sugar pig snatched from the tray of an itinerant hawker. But there is no denying the sad fates of the Little Piggies who "had none" and were sent "crying all the way home," or the tragic deaths of the two imprudent Little Pigs who built their homes out of straw or sticks.

There have been recent ill-advised attempts to produce more politically correct versions of classical folktales such as the *Three Little Pigs,* in which all three pigs learn their lessons, take out proper wolf insurance, and live happily ever after. I am pleased to report that even very young children will have none of such nonsense. There is an essential, even blood-thirsty streak of realism in preliterate children, who already understand the way the world works and expect a certain amount of attrition. They want to see justice done and take huge pleasure in the successful strategies of the third Little Pig. They prefer a minimum of caution in their cautionary tales.

We do well to remember also that literary pigs are images dressed up in human clothing, with human faces and largely human concerns. They are the only pigs most city children know and they bear no resemblance to real animals, but they are accepted as such very readily by children who have a quick understanding of the conventions used in picture books—and enjoy them anyway. The best children's books work within these guidelines and produce a captivating mix of fantasy and reality.

Beatrix Potter was perhaps the first to use the technique deliberately. The *Tale of Pigling Bland* was published in 1913, the year she married and left home, in spite of her parents' objections, to live on Hill Top Farm in the Lake District. The story follows Pigling, a lovable inno-cent, partly gauche and tactless, partly dutiful and priggish, with a romantic streak. His dream is "to have a little garden and grow potatoes," a dream Potter shared and illustrated with the view from her own garden at Hill Top. Pigling meets Pig-wig, "a perfectly lovely little black Berkshire pig with twinkly little screwed-up eyes, a double chin, and a short turned-up nose." Together they escape, "over the hills and far away," in a tender love story of great sensitivity. Graham Greene credits Potter with being "the first to throw any real light on the Love Life of the Pig," which she did "with delicacy and a psychological insight that recalls Jane Austen." The characters are rich and affectionately drawn, in pictures and words that have become the envy of many serious novelists and the delight of generations of children. And all this is accomplished without losing the essential piggishness of the two main characters, who would not have been nearly so attractive had they been portrayed as dogs or monkeys.

Almost twenty years later, in 1930, Potter returned with the *Tale of Little Pig Robinson,* who is marooned on a desert island, but it was not nearly as successful, as a pig or as a story. The hero is greedy, callous, and obtuse and might just as easily have been depicted as a baboon. Potter's best writing was an elegant and unsentimental way of escaping from a lonely and repressed childhood, and what was an escape for her was also a faithful and keenly observed act of animal liberation that has given pleasure to pig lovers and mere humans alike.

In the meantime, A. A. Milne in 1926 produced *Winnie-the-Pooh* and the charming Piglet, who was chronically nervous, filled with anxiety about being "A Very Small Animal Entirely Surrounded by Water" and with "nothing to do until Friday." He was a fitting companion to Christopher Robin and his friends in the Hundred Acre

SAINT ANTHONY, PATRON OF swineherds, is usually portrayed with a plump pig at his feet.

Wood, but he was not really a pig at all, though he was honored later by Benjamin Hoff, who published a spoof New Age tribute called the *Te of Piglet*.

Winnie-the-Pooh was followed in 1927 by *Freddy Goes to Florida*, the first of twenty-six books by Walter Brooks, a staff writer at the *New Yorker*. Freddy is a smart aleck, "the youngest and cleverest of pigs," who goes on adventures with a barnyard of friends and solves impossible mysteries right through until 1958 without getting a day older. He was also the first pig to be "outed." In the prescient *Freddy Goes Camping* of 1933, Freddy is portrayed by the artist Kurt Wiese trotting down a wide staircase in a billowing floral frock and wide straw hat.

The next pig of any consequence arrived on the scene in 1952, courtesy of E. B. White, also a contributing editor at the *New Yorker,* with his classic children's book *Charlotte's Web*. This was sparked by White's own experience of seeing the slaughter of a pet pig and features Wilbur, a hero pig, saved from a similar fate by the wisdom of a cosmic spider called Charlotte. Wilbur is a real pig, who digs himself a warm tunnel in the straw of the barn and eats "skim milk,

BABE, THE HEROIC LITTLE GILT, became a sheep-pig to save the farmer's face and his own bacon.

wheat midlings, leftover pancakes, half a doughnut, the rind of a summer squash, two pieces of stale toast, a third of a ginger snap, a fish tail, one orange peel, several noodles from a noodle soup, the scum off a cup of cocoa, an ancient jelly-roll, and a spoonful of raspberry jello"—perhaps the best and most accurate description ever of the diet of a cottage pig.

White succeeded wonderfully in telling a fantasy tale without distorting the character and real natural behavior of pigs, though he insisted that it was not a moral tale, because animals are essentially amoral. He described it instead as "an appreciative story which celebrates life, the seasons, the goodness of the barn, the beauty of the world, and the glory of everything."

Wilbur was a hard act to follow, and no subsequent fictional pig in a children's book has ever quite matched up to him. But, while pigs have been given a mixed reception on film, several celluloid pigs have come close.

The first pig in a Hollywood production was a prize hog called Blue Boy, who was brought to the Iowa State Fair by the Frake family, played by Janet Gaynor and Will Rogers, in the 1939 version of *State Fair*. There was also a Rogers and Hammerstein musical version with

Jeanne Crain and Dana Andrews in 1945; and, for reasons hard to comprehend, a 1962 remake with Pat Boone and Pamela Tiffen, in which Tom Ewell sings to the pig.

The absurd side of pig mania surfaced in 1969, when Tom O'Horgan somehow managed to find the funds to turn his off-Broadway play about a man who marries his pet pig Amanda into a feature film with the prophetic title of *Futz*. And in 1984, the dark side of boarness was exploited in a ludicrous horror film called *Razorback*, about a giant pig that terrorizes the Australian countryside—probably the worst thing that ever happened to a nice animal.

In 1985, film went some way toward making amends with *A Private Function*, an Alan Bennett story set in food-rationing times after the Second World War, in which a podiatrist and his starchy wife try to conceal a contraband large white pig called Betty in their semi-detached home. Though the human lead roles were played by Michael Palin and Maggie Smith, it was the pig who stole the show.

Cinema hasn't really done pigs justice, though in 1994 a small British film called variously *Hour of the Pig* or *The Advocate* did explore some of the less obvious possibilities. In it Colin Firth, as a Parisian lawyer in the fifteenth century, finds himself on a murder case in the country defending a pig—a courtroom drama with a difference. But the runaway box-office success of 1995, winning an Oscar for special effects, was *Babe*, adapted from Dick King-Smith's story of the sheep-pig. This could so easily have been awful sentimental swill, but the Australian director Chris Noonan captured pig appeal so perfectly that a succession of little white gilts, with computer assistance, produced a delightful and disarming film.

Pigs have also been a staple of cartoons. Beginning in the 1930s and continuing right into the 1960s, Warner Brothers introduced audiences to a round, pink, good-natured Yorkshire barrow called Porky Pig, who was so sanitized he had no genitals. Porky spent most of his time chasing, but never catching, a pesky rabbit called Bugs Bunny. His cartoons were all directed by Fritz Freleng and his characteristic stammer was provided by Mel Blanc. But it was the pig who always had the last word—"That's all, folks."

More recently, another successful animated pig came from Disney. The sometimes dark film *The Lion King* was leavened with the comic coupling of a fast-talking, wild-wise meerkat called Timon, wonderfully voiced by Nathan Lane, an his unlikely straightman Pumbaa, a simple, honest, but flatulent warthog, dubbed by the versatile Ernie Sabella. Together, they put a gang of menacing hyena to flight with the transformation of the gentle warthog into a heroic hog who announces: "They call me MISTER Pig!"

There have been plenty of other more or less successful cartoon pigs, but the all-time queen of animated anthropomorphic pigs has to be the outrageous Miss Piggy from Jim Henson's Muppets, a wild collection of characters who stormed into our lives on television and in feature films such as *The Muppet Movie, The Muppets Take Manhattan,* and *Muppet Treasure Island.*

In the earliest episodes, Miss Piggy was just one of the cast, but she was soon plucked from the chorus and found fame around the guiding hand of Frank Oz. He showcased her talents with wonderful extravagance, allowing the blonde-wigged, blue-eyed *cochon fatale* to go over the top in every way, except when it came to her bashful, blushing, and unrequited love

MISS PIGGY, HOGGING THE limelight as usual.

for Kermit the Frog. Initially, Miss Piggy wore a gray lounge-singer's dress, but later she acquired a complete wardrobe. She has now worn costumes in the style of all the Hollywood glamour queens and has been honored by special shows featuring "Miss Piggy's Treasury of Art Masterpieces from the Kermitage Collection." She may well be the greatest animal celebrity outside Disney's control, a phenomenon everywhere except the Middle East, where her fame is a source of uneasy bewilderment.

LONG BEFORE MOVING PICTURES, HOWEVER, ARTISTS AND SCULPTORS WERE DRAWING ON PIGS for inspiration. The oldest known illustration of a pig, painted about 40,000 years ago in the cave of Altamira in northern Spain, depicts a wild boar in the act of leaping to the attack, with his forelimbs off the ground. It is a representation with great presence, making it easy to believe that the artist had more than just a local animal or a single event in mind. The result is a finely observed record of natural history, and much more—an evocation of magical power perhaps, an act of propitiation, or a prayer that touches on the spirit of the hunt.

This votive approach continued into early historical times. Greek pottery painters of the first millennium decorated their amphorae and plates with equally rampant wild boar, some in full stride and each crisply rendered to show every bristle and tusk. Others clearly celebrate the capture of the Erymanthian boar by Hercules, or the defeat of the Calydonian boar by the great huntress Atlanta.

The first portrayals of domestic pigs are probably Chinese offerings found in Zhou tombs of the ninth century BC. There is an Assyrian frieze from the seventh century BC, now in the Louvre, that features a sow followed by her pigs, and several early Egyptian representations of sows and piglets connected with the worship of Isis, the chief goddess, or her mother Nut. There are also some charming works by the Greek artist known only as "the Pig Painter," who decorated several vases with exquisite scenes of Eumaios, the swineherd of Odysseus, and his charges in the fifth century BC, while in the Vatican Museum there is a life-size marble sculpture of the "Great White Sow" that Aeneas found marking the site chosen for the new city of Rome. She was honored again in the first century AD by having her likeness minted on the *sestertius* coin by the Emperor Antoninus Pius.

It is often difficult to distinguish some early domestic pigs from their ancestral boars, but by the third or fourth century AD, differences are obvious. Jade pigs from China, ebony pigs from Egypt, pigs on gems and seals and precious stones from India, leather puppet pigs from Java, and pigs carved from horn and ivory in Europe, all begin to appear in markets everywhere, a clear sign of the global fascination with pig images and piggishness. Particularly popular still are pigs in terracotta or porcelain. Indeed, the very word is derived from *porcella,* meaning a "young sow," while *porcelaine* in French also refers to the cowrie shell, whose smooth, enamel texture and shape resemble a pig's back.

Today, the most common pig in pottery is the "money pig" or "piggy bank," a pig-shaped piece of pottery with a coin slot on its back, designed to collect small change that can only be released by smashing the whole container. It is said that piggy banks originated in twelfth-century China as burial goods, pig-shaped pottery filled with gold for the next world. But Western tradition has it that this kind of money-box was invented by the seventeenth-century French engineer Sebastian le Prestre, who calculated that one pig in ten years could multiply to more than 6 million offspring. He decided that there was no better model for teaching the young to save than a prudent and prolific pig.

In the nineteenth century, Cadborough Pottery in Sussex produced a pig with a less instructive but perhaps more entertaining purpose—a thirsty-looking porcelain pig whose head could be detached to form a drinking-cup for some cheering liquor housed in the body of the beast. At Sussex weddings, it was customary to drink to the bride's health in this way, so that everyone involved could claim to have taken a "hogshead" in her honor.

PIGS AND PORCELAIN: AN ancient potsherd and a modern piggy bank.

ALBRECHT DÜRER'S COPPERPLATE engraving shows the Prodigal Son at the low point of his exile as a swineherd.

It is perhaps no more than a nice coincidence, but the first painter of note to produce works whose main subjects were pigs was Paulus Potter, a minor Dutch master who was celebrated in the mid-seventeenth century for his landscapes that caught animals in harmony with the passing moods of nature. Each composition came complete with pigs, swineherd, scruffy trees, and perhaps a dog or a wagon wheel, all looking very much at home in their rustic sur-roundings. His direct and conscious heir was George Morland who, in the eighteenth century, made pencil drawings and etchings that catch every expression in a pig's eye: "The pertness, the liveliness, the humour, the love of mischief, the fiendish ingenuity and perversity of which pigs are capable, can be fully known to the careworn pig-minder alone."

Morland may have seen the pig drawings of Dürer, Rubens, and Rembrandt, but he learned on the job, exhibiting at the Royal Academy when he was ten. He picked up anatomy by slaughtering and, from George Stubbs's *Anatomy of the Horse,* rode as an amateur jockey, and he flouted lucrative commissioned work in favor of sketching in the country from life.

Many artists have tackled pigs for versions of the tale of the Prodigal Son, with mixed results. Breughel, both Elder and Younger, and Millet included pigs in their landscapes, and more recently, Charles Tunnicliffe and Stanley Spencer have dabbled in pigs, as has Marc Chagall, who made his green. Winslow Homer painted a pair of watercolor pigs looking very much at home in a glade on Bermuda, and James Wyeth created an ecstatic life-size portrait of a 450-pound Yorkshire sow called Den-Den, who ate twelve tubes of oil paint during her sitting. "I became enamored of her," he said. "There were good vibrations between us. Her eyes were so human, just like a Kennedy's." But, as a draftsman of pigs, I don't think Morland has ever been surpassed. His hogs do everything but squeal.

On the commercial side of painted pigs, some of the best work can be found on Britain's pub signs. The older the pub the better, because the inns with some history still have signs that are heraldically correct. The Boar's Head, The Boar's Arms, and The Blue Boar abound and are self-explanatory, as are The Little Pig, The Butcher's Arms, and The Hampshire Hog, all

illustrated in bright paint and emblazoned with an appropriate coat of arms. But what is one to make of The Pig and Whistle?

There are medieval church carvings that depict pigs playing pipes, but that does not seem to be the origin. In Scotland, they say that a "pig" is an earthenware pot for dispensing cheap whisky, and that "whistle" in the Highlands means small change or petty cash. That sounds plausible, but in the Borders they claim that *Pige-Washael* is an ancient salutation to Our Lady. In Northern Ireland, a pub called The Pig and Whistle explains that pig-drivers used to drink there while waiting for the livestock train to arrive. Perhaps the name simply exists because someone liked the sound of it, or because there is an arcane meaning behind it. There were once almost as many White Boars, all bearing the arms and standard of Richard III. But when that ill-fated monarch was killed at the Battle of Bosworth, the innkeepers painted their white pigs blue, hoping to appease Henry VII with the mark and sign of his champion, the earl of Oxford, who was always "true blue."

IN 1983 THE ARTIST AND AUTHOR MICHAEL RYBA PUBLISHED A GROUND-BREAKING WORK ENTITLED *The Pig in Art,* in which he argues persuasively for a totally new view of the history of art. It has, he suggests, ever since the birth of human awareness, been driven by a secret cult that regards pigs with religious veneration.

In evidence, he offers a Stone Age figurine known as the "Venus of Willendorf," which does have distinctly porcine proportions; a number of Greek vases in the series called the "Attic Travesty," in which pigs are portrayed in vigorous and carefree attitudes; the insight that Dürer's grandfather was a butcher in explanation for the many pig motifs in his engravings; an early study of the Mona Lisa that attributes her secret smile to the fact that beneath her flowing robe a trotter is distinctly visible; and Peter Paul Rubens's lost masterpiece "The Rape of the Sabine Sows," in which the figure seen in silhouette in the foreground may not be his muse but rather a poetic pig. Whatever the case, it is hard not to agree with Ryba that "the touching appeal of the pig, an animal too often underestimated as of purely practical use, should not go unheeded by the artist of the present day, nor those of generations to come." And it is impossible to ignore the wry appeal of his illustrations.

Pigs have, in truth, become transcendent symbols in human culture. They have, at one time or another, come to mean all things to all men. Pigs are simultaneously the objects of disgust and adoration. They have been made the source of all evil by the same people who once stirred reverently through pigs' entrails looking for signs from the gods. The Romans even had a word for it—"haruspication." Others did, and may still, look to pigs even for forecasts of the weather. "Pigs see the wind," said Samuel Butler, knowingly.

The Greeks were obsessed by pigs, even in pre-Hellenic times when they worshiped a gray Crone-Goddess who was depicted as a sow. Joseph Campbell describes the sacrifices to her as "a holocaust of pigs—carried out with a certain element of chilly gloom." And even when worship was directed by less saturnine Olympian gods, a large number of deities continued to be suckled or slain by a pig of some kind.

The great anthropologist James Frazer describes the presence of pigs in ancient mythology as "pervasive," and his retelling of the story of Persephone's seduction and her disappearance into the underworld concentrates on the pigs that led her there, suggesting that these were the embodiment of her mother Demeter, the corn goddess. So the tracks Persephone follows were those of her mother and herself—in effect "The pigs *were* Persephone." And so it goes, throughout history. The image and the influence of the pig are truly ubiquitous.

A PIGLET DEEP IN SWILL ON A pub sign in Britain.

The wild boar was associated in Greece with gods and heroes, revered for its ferocity and strength and used as the symbol of Artemis and her capacity to unleash sudden violent destruction. Circe turns Odysseus's men into pigs, not dogs. Plato even devoted one of his Socratic dialogues to swineherds. And it was pigs that gave Rome its "Pyrrhic victory" over Carthage by stampeding Hannibal's elephants.

Imperial tombs in the Han dynasty were embellished with life-size ceramic pigs long before they featured terracotta armies of human protectors. When devils are cast out, they flee in the bodies of Gadarene swine, not cats or rats, while the legendary banquet of Trimalchio is devoted almost entirely to whole hogs, blood puddings, and pork sausages. Galen, the Roman physician, prevented from dissecting humans, chose pigs as his subjects and correctly identified the functions of the kidneys, bladder, heart, and spinal cord. The Emperor Heliogabalus rode through the streets of Rome in a pig-drawn chariot. And in Hindu

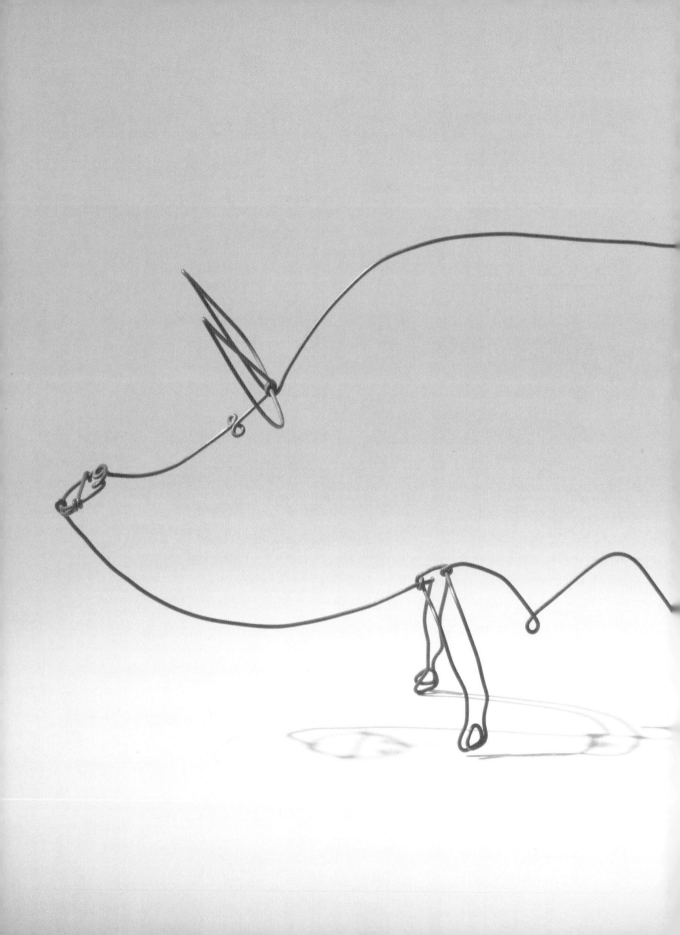

iconology, Brahma, Vishnu, and Siva are all carried through the clouds in chariots drawn by hogs.

Pigs infiltrated culture at every level, everywhere. Irish mythology maintains that the first to arrive on the island was Bamba, one of the daughters of Cain, whose name means "pig." Celtic graves and Anglo-Saxon armor were decorated with boarish motifs and, at about the same time, the Twentieth Roman Legion in Britain adopted the wild boar as its emblem. Right through into medieval times, art continued to both canonize and anathematize pigs. Genre painters memorialized prize pigs and their owners; Fabergé, in the nineteenth century, made pigs of pink rhodolite with diamond eyes; and folk artists throughout Europe still translate the roundness of pigs into homely objects such as jugs and weather vanes.

The message that runs through all these representations is the same. "Nothing is quite what it seems." There is more to pigs than meets the eye. Arthur Rackham in 1901 illustrated *Alice in Wonderland* as a romantic waif in a sedge garden with an adorable lace-capped piglet in her arms. Alexander Calder in 1928 created a whirly sow in wire for the Museum of Modern Art in New York. Lord Snowdon produced a sexy Yorkshire with silky buttocks for the British edition of *Vogue* magazine in 1990. And James Dean was memorably photographed on his uncle's farm in Indiana in 1955, wearing an aviator jacket and jeans and carrying a cloth cap in one hand, while the other gloved hand rests proprietorialy on the back of a morose Chester white pig. His stance suggests that of a hunter with his trophy, all very "ham."

Pigs are great hams. Nothing else in the farmyard can keep pace with them. They learn to operate automatic feeding systems after a single demonstration. Most cows fail to figure it out and starve instead, and horses never stop eating, sometimes ending in an explosive death. But pigs take just what they need and put on 1 pound of flesh for every 3 pounds of corn. Cows require 10 pounds of grain to show the same improvement. Perhaps that accounts for the fact that the oldest continuously operating academic club in America is Harvard University's Porcellian Club, founded in 1794 and still the most exclusive and influential association of achievers in the world.

Nobody is neutral about pigs. On one side are the pig haters afflicted with "hogrophobia," who see them as dirty, greedy, disgusting beasts too evil to touch, let alone eat. On the other side are those who find pigs to be fastidious, captivating, highly intelligent creatures whose very corpulence is appealing. The "pig haters" are catered to by political cartoons in which pigs are portrayed as lethargic and loutish gluttons, bloated swine feeding on the public trough, while the "pig lovers" are entertained by perky, pink, well-scrubbed little pigs, alive with liberal intellectual energy. This division seems superficially to be one between hard masculine and soft feminine options, but that doesn't quite identify the difference. Those who choose to work with pigs in America make much of the fact that they are known as hogmen, while their counterparts who tend to cattle and strut around in chaps and silver spurs are known as cowboys.

PIGS DISPORT THEMSELVES IN EVERY LANGUAGE. LIVING "HIGH ON THE HOG" HAS ITS ORIGINS IN the nineteenth-century British Army custom of giving pigs' feet to the enlisted men but saving the hams and shoulders for officers. "Pork-barrel politics" were once practiced by politicians who traded gifts for votes, but the phrase now describes the behavior of members of the US Congress who push for federal projects in their own home districts. "Buying a pig in a poke" is risky. "Casting pearls before swine" is wasteful. "Carrying pigs to market" is businesslike, but "bringing one's pig to the wrong market" is just a bad deal. "Pigs with wings" are highly unlikely. And while it could be a compliment to call a Harley Davidson a "hog," it is downright rude to the animal to associate male chauvinists with pigs.

In French, *une histoire cochon* is a smutty story. In Italian, *cani e porci* is a very pithy description of a gang of worthless layabouts. In German, *kein Schwein etwas gesagt* indicates a gross lack of interest in any subject. These all betray a certain discourteous and unmannerly attitude toward pigs as a whole, but whatever your feelings about pigs may be, there is no denying that they are highly symbolic and even capable of having a foot on both sides of the metaphoric divide.

For instance, Osiris, the Egyptian god of the dead and afterlife, had a pig sacrificed to him on the anniversary of his death. Osiris was murdered by Set, who is often depicted with a pig's head, wielding a flint knife as he sets about dismembering his brother. The pig is a symbol for them both, a contradiction that James Frazer glosses over with the assurance that "the god is sacrificed to himself on the ground that he is his own enemy"—a wonderfully slippery description of the complex nature of the pig.

In most studies of symbolism, the pig represents gluttony and sensuality. It wallows in mud, demonstrating its insatiable appetite for sloth, selfishness, lust, and greed without remorse. Saint Clement of Alexandria warned that the eating of pork therefore would make one fat and swinish, liable to take on all the lewd and greedy characteristics attributed to the pig. But pigs were also associated with fertility and believed capable of bringing rain. In Egypt, people once even wore pig amulets.

Pigs earned the love of Demeter the Earth Mother for their fertility. She was often portrayed with a pig cradled in her arms, but paradoxically she also seemed to require pig sacrifices in her name, perhaps a reflection of the fact that pigs are known, on occasion, to devour their own young. They also eat human corpses and were, on that account, regarded as "unclean" by Egyptians, Hebrews, Muslims, and Phoenicians. But that didn't prevent the Egyptians, at least, from sacrificing a pig each year to Osiris and enjoying its meat. The unlawful, it seems, can be lawful on special occasions.

Leprosy was once blamed on pig's milk, a notion that may explain why swineherds were excluded from Egyptian temples. But even they were sometimes granted special dispensation in the light of persistent folktales that tell of princesses forced to marry swineherds, only to discover that these are really princes in disguise. Some people who refuse to eat pork maintain that the strange dark marks that exist on the inner side of a pig's forelegs are blemishes left by the devil, who entered the animal there. You can see these scars quite clearly when a dead pig is shaved, but they have no apparent anatomical function.

Ambrose Bierce, the American writer with a dark, sardonic eye, nicely encapsulated all this confusion in his *Devil's Dictionary*. His ironic definition of the word "edible" is "good to eat, and wholesome to digest, as a worm to a toad, a snake to a pig, a pig to a man, and a man to a worm."

In other hands, pigs are seen to be lucky. A Teutonic tradition holds that pigs bring good luck at the New Year and that white sows are particularly fortunate, especially in the matter of money. Such pigs are often seen with a four-leafed clover in their mouths or standing next to a bag of gold coins—hence the proverbial "pig in clover," which signifies financial prosperity and well-being. In Germany it was once the custom to award a piglet to the loser in a race, as a consolation prize.

In Latvia, planting a pig's tail helps to ensure a good barley harvest. Welsh folklore holds that Saint Elmo's Fire can only be discharged by the squeal of a pig. The Irish say that meeting a sow with a litter brings good fortune to a bridal procession, and walking around a pigsty three times in a clockwise direction can cure most illnesses. The boar at the hub of the Buddhist "wheel of existence" signifies passion. Pigs carved on the columns of a Shinto shrine in Japan symbolize courage. And in China, pigs are associated with virility.

Those born in the Chinese Year of the Pig—such as 1935, 1947, 1959, 1971, 1983, and 1995—are likely to be good-natured, honest, and reliable. They work hard, play hard, and are generous to a fault. Their weaknesses are all connected with affairs of the heart, in which they are maladept. The list of famous "pigs" includes Al Capone, Elton John, Ronald Reagan, Ernest Hemingway, Humphrey Bogart, and Alfred Hitchcock, who once said: "I understand the inventor of the bagpipes was inspired when he saw a man carrying an indignant, asthmatic pig under his arm. Unfortunately, the man-made sound never equalled the purity of the sound achieved by the pig."

Our love/hate relationship with pigs continues. It is part of our fascination with them, and our uneasy awareness of their regard. There is a bond between us that runs like a thread right through history, coupling us in ceremonies that make pigs and people interchangeable. And yet we still hold them at arm's length, afraid of getting too close and seeing ourselves too clearly in the mirror they provide. But there is no avoiding the surprising conclusion that no other two species, from such different origins, have so much in common.

WARTY PIG

The warty pigs belong to the same genus as the wild boar and are descended from an Asian lineage, which may still survive in the forests of Vietnam.

But those that live on volcanic outcrops along the Wallace Line on the Sunda continental shelf are all variations of the same theme.

These species of *Sus* look superficially alike and share the same antediluvian appearance. They have an awkward, thrown-together quality, like something stitched up by a sorcerer's apprentice. Their ordinary barrel-shaped bodies are balanced on borrowed limbs and topped by heads that resemble grotesque Mardi Gras masks. The ears are small, the faces flat, and the muzzles extruded and square, which makes them look disturbingly like horizontal orangutans.

They are relatively small, as island races tend to be, and most have sparse coverings of short bristles on their cheeks and chins that blossom into Edwardian sideburns and beards. The boars are more than twice the size of their mates and have large curved tusks on both jaws. Their tails are long and simply tufted at the tip.

The islands on which warty pigs are found are not far apart. You can travel all the way from the Pacific to the Indian Ocean without ever being out of sight of one or another of the island volcanic cones that rest on a shelf less than 400 feet deep. There have been times during the last 10 million years when the islands were connected

by dry land-bridges, but even if they missed such opportunities, all these pigs are strong swimmers. They wallow in the same gene pool and share features that allow all kinds of relationships to be drawn between them, but it is still possible to identify at least five more or less distinct warty pig species by examining the pattern of male facial adornments.

These "warts" fall into three convenient diagnostic areas. They gather as *infraorbital* warts on the cheek swellings, *preorbital* warts directly above the upper canine teeth, and *mandibular* warts on the angle of the

THE JAVANESE WARTY pig *Sus verrucosus* with a facial shield of three pairs of outgrowths.

jaw. Each species has a diagnostic pattern.

Starting in the north, *Sus philippinus,* the Philippine warty pig, occurs on all the large islands of Luzon, Samar, Mindoro, and Mindanao. This species has just two pairs of warts, large preorbital and small mandibular ones. It also has a crown of unruly black bristles arranged like a bad toupé, a full mane down the neck, and white tufts on the angle of the jaw.

There is a second species on the islands of Panay and Negros. *Sus cebifrons,* from the Latin *cebus* and *frons* for "pale-browed," or the Visayan warty pig, is a rare and small species with very small warts on the same two sites. It is dark-colored and has a distinctive whitish band around the snout, and a long mane.

Farther south, in Indonesian waters and in the Moluccas, there is *Sus celebensis,* the Sulawesi warty pig, which has three pairs of warts. The preorbital ones are very large, the infraorbitals a little smaller, and the mandibular almost entirely lost from sight in a whorl of long hair. This species has some reddish hair on the coat, a crest of very long hair on the crown, and a distinctive yellow snout band.

Sus verrucosus, from the Latin *verruca* for a "wart," is the Javan warty pig, confined to the large island of Java and the small offshore island of Bawean. This is the most typical of the warty pigs, with three pairs of the outgrowths on every male, but in this case the preorbital ones are less well developed. The coat is generally dark, but the underparts are yellow and strongly demarcated. A long-haired mane covers the neck and tails away down the flat back. The Javan warty pig is endangered.

The fifth species is *Sus barbatus,* from the Latin for "bearded," and is known as the bearded pig. Despite its name, this is a warty pig, with both preorbital and infraorbital warts well marked but largely concealed beneath a lavish set of gray whiskers. Even the females are bearded. This is a large pig with an extraordinarily long snout. The body hair is short and bristly. It is more widespread than the other four warty pigs, occurring still in some numbers throughout Borneo, Sumatra, parts of the Philippines, and on peninsular Malaysia. Bearded pigs, like white-lipped peccaries, are devoted to the oil-rich seeds of oaks, chestnuts, and members of the dipterocarp family, which includes timber trees such as camphor and copal. All these broad-leafed trees provide mast in their season, and when this fails, bearded pigs manage well enough on the flowers and fruits of other species, plus worms and mushrooms and roots. Every five years or so, however, the dipterocarps go crazy, fruit heavily, and produce bumper crops of thousands of tons of seeds, and the bearded ones feast. Sometimes they follow troops of macaques and gibbons, but most often they go their own way, making great migrations, following wide, well-worn paths that have been used for centuries, traveling by night and resting up in thickets during the day. Such bands include a hundred animals or more, collectively less shy and more pushy than the usual family groups. And from time to time (the last time was in 1987), they aggregate in extraordinary population eruptions.

These concentrated herds, sometimes thousands strong, swarm even in daylight, pouring through the forest, sending pigs of all sizes charging through the undergrowth on a broad front, making their own roads, fording rivers, crossing coastal bays, or even swimming out to sea. They seem to travel consistently in one direction over a period of weeks, and those that survive retrace the same route back or follow the species memory of a grand circular course to return to where they began.

Zoologists lucky enough to witness this phenomenon suggest that the stampede is connected to synchronized flowering and mast-fruiting of their favorite foods, but this is not always obvious. What is certain is that many of the migrants die, just as lemmings do, and perhaps for the same reason. This may be a natural cyclic cull of populations that have got too large for their own good. In Kalimantan in 1954, so many bearded pigs swept across the Kayan River that Dyak hunters were able to massacre tens of thousands of them, and so many carcasses drifted down-river that deeply offended Muslim communities on the coast declared war on the perpetrators upstream.

IN THE NIAH CAVES OF SARAWAK, ARCHAEOLOGISTS HAVE FOUND LARGE NUMBERS OF PIG BONES and teeth together with human remains 40,000 years old. Most of the pig bones were butchered bearded pigs, and this one species still provides most of the protein consumed by Sarawak's 1.5

million people—over 32,000 tons of carcass every year. Pork is not "the other white meat" here; it is almost the only meat in most markets.

It has always been that way in the islands. In cave middens excavated on Sulawesi, the main item on the menu was the Sulawesi warty pig. And because this was so, it has been suggested that this species may have been domesticated a long time ago. That means that domestication of wild pigs could have taken place in these islands before, or at least at the same time as, the wild boar, *Sus scrofa,* was being raised on the mainland of Asia.

The evidence for this rival agricultural enterprise comes mainly from the discovery of *Sus celebensis* bones on Flores and Timor and in the Moluccas, bones showing the usual trademark changes that take place during domestication. That in turn implies that such

THE MOST INTERESTING FEATURE of the bearded pig is its occasional lemminglike migrations. No other pig migrates.

pigs were traded at these sites or taken there by settlers from Sulawesi at least 5,000 years ago. Allowing another millennium or two for domestication, this puts the original capture and breeding of wild pigs on Sulawesi into the same time-frame as the pioneering activities taking place in Neolithic China.

The two centers may even have made contact with one another surprisingly early. The Chinese were certainly trading by land and sea by 3000 BC and would not have missed the opportunity to exchange pigs for pearls and spices in the islands. And even without that exchange, there was nothing to stop the boar, wild or domesticated, from swimming under its own steam out along the island chain from the Malay peninsula. There is plenty of evidence of *Sus scrofa* in the islands, some of it prehistoric.

It is interesting to note that the second species of wild *Sus* ever to be described was *Sus papuensis,* found on and considered to be native to New Guinea by scientists on the French exploration ship *La Coquille* in 1826. They saw this pig domesticated and running wild around almost every village on the island and it never occurred to them that it was not native to New Guinea.

It never occurred to zoologists during the following century, either. But eventually all the pigs of Melanesia began to look uncomfortably out of place and they were described as

"The Papuensis Problem." The mystery wasn't solved until 1981, when Colin Groves from the Australian National University published a meticulous taxonomic study of all the known species in the genus *Sus*. His multivalent measurements show that *Sus papuensis* never existed as a species in its own right. It is a hybrid between *Sus scrofa* and *Sus celebensis*—in other words, between the only two wild pigs ever to be domesticated anywhere.

IN THE EARLY 1970S, I HAD THE INTERESTING MISFORTUNE TO BE CAUGHT IN AN UNUSUAL CYCLONE in the Banda Sea, east of Bali. I was traveling in a tiny wooden prau with two Javanese crewmen, and we washed ashore on a small island in the Lesser Sundas.

I spent nine months among friendly people, living in a tiny, two-room hut with walls of woven cane. The roof was thatched with palm leaves, and the floor was slatted driftwood raised a few feet off the ground to keep it cool. My door opened on to a small veranda decorated with wild orchids growing in bamboo tubes, and looked out across the lagoon toward several small islets of raised coral, fringed with patches of green mangrove. It was idyllic, and to complete my delight, I discovered that my house came fully furnished with a resident pig.

Her name was Pig, which seemed uninspired, but it sounded better in Indonesian. She answered to Babi. And Babi was a charmer.

She was fully grown, about 2 feet tall and 3 feet long, weighing I guessed about 60 sturdy pounds. Her coat was black, brightened by red and white tufts and a striking yellow band that encircled her snout. Being a female, her three facial warts were small and unobtrusive, but on her head she wore a tuft of long bristles that flounced like the feathered headdress of a dancer at the Moulin Rouge. She was a proto-punk pig.

Babi lived under my floor. She had scraped out a hollow in the sand there and lined it with dry leaves and grass to make a comfortable nest from which she could see anyone coming up the path. She was a watch-pig, the best guard and lookout I have ever known, made vigilant by the fact that she was an outsider herself. As the only pig in a Muslim village, she was used to being alien and strange, an object of suspicion.

We hit it off right away. She was delighted to have company and reassured to discover that I looked, sounded, and smelled totally unlike any of the villagers, who tended to give her a wide berth. They were, after all, forbidden by Islamic law to have anything to do with pigs, but Muslims in Indonesia were then reasonably relaxed about almost everything, and I soon discovered that some of the old women had been leaving morsels out for Babi. I never did discover how she had come to be there in the first place, but I soon found that she was perfectly capable of looking after herself.

We walked the beach together at low tide, combing the strand for all the usual jetsam of fish, jellyfish, starfish, and sea cucumbers. She was particularly fond of mussels and limpets, which she prized off the rocks with her sharp canine teeth and chewed on with a lot of lip-smacking and sucking. Oysters left her cold. Perhaps she found them too messy, because she was the most fastidious eater. The only exception was the night she found a hoard of palm wine hidden in the forest out of sight of the imam. Then she behaved like any other proverbial pig and came home disgustingly drunk.

Babi was the only pig on that island, but I gathered that there were feral pigs on some of the larger islands nearby, and one old fisherman told me that he had seen pigs like her in the markets of Sulawesi. "They are infidels there," he said, "and know no better, but it is true that such an animal may bring as much as 10,000 rupiahs."

I was able later to visit Minahasa on the long, northern arm of Sulawesi and to confirm that Babi was indeed a typical Sulawesi warty pig of the species *Sus celebensis,* and that hundreds like her are being hunted and sold every year. But very little is known about their ecology or behavior. Thanks to Babi, I can fill some of those gaps.

To start with, Celebes or Sulawesi warty pigs are diurnal and intensely social. Babi came out of her nest every morning an hour before sunrise. I know, because she used to come up the stairs to my door and make a single husky "hah" sound, a short, clipped exhalation that combined greeting and reprimand. "Hey! Not up yet?" Then she would stand there, looking impatient, making a repeated plaintive "wheeh" sound, which I recognized as mild irritation.

Every day was different, but she made the important decisions. First light always found us out on the beach or on a path that led instead to one of her favorite inland haunts. She set the pace and made sure that I kept up by giving an encouraging little grunt every now and then—the sort of sound that pig-ignorant people describe as "oink" and presume to be the only vocal expression of which pigs are capable. It was obviously a contact call just loud enough for me to hear, without attracting the unwanted attention of possible predators. And it was one of the few communications that required an answer. If I failed to reply with a sound of my own, she would stop, turn and "wheeh" me once or twice in exasperation.

Sometimes we went up the volcano slope to check on banyan trees and assess the ripeness of their figlike fruits. At other times, she introduced me to cool fissures in the rock where mushrooms could be found. On one occasion, I picked an interesting bracket fungus growing on a dead stump and she made an explosive "woof" sound at me, going on until I put it down. I followed her guidance on fungi after that. It is a chastening experience to be barked at by a pig who objects to your table manners.

My favorite warty pig sound is a querulous "auk" on a rising tone. It communicates surprise. Babi made it when she found anything new and curious, such as a chameleon, or when she heard a sound she could not place. It was always worth pursuing. I treasure the memory of digging one day into a burrow on the beach that I knew would expose a large ghost crab seeking shelter from the midday sun. Babi came over a little dubiously and sat back on her haunches to watch me at work. But when I reached the crab's subterranean chamber, its occupant burst out as though it were spring-loaded and skidded down to the water, moving so fast that neither of us had any chance of catching it. All Babi had time for was a high-pitched "auk" and a little see-saw movement of her head that I could swear was pig laughter.

There was much more to her vocal repertoire. A burp of sheer pleasure when feeding on something special. A "wonk," which meant "leave something for me." A loud alarm call that sounded like an ambulance and doubled as a warning. And, just once, when a heavy branch fell on her back, an ear-splitting scream that brought me running out of school to see what had happened.

Babi loved to be groomed. Anywhere on her neck or ears was heaven. It turned her legs to jelly and would end with her leaning heavily on me with her eyes closed. And if I moved, she would roll right over with all four trotters in the air, waiting to have her belly rubbed. To me this seemed obvious evidence that warty pigs can and do provide for each other whenever they are together in secure resting places. I would be willing to bet that they overnight in heaps and snore each other into dream sleep. Babi did just that some nights, rattling the shutters on my window.

But the thing I found most endearing about this pig was her very real pleasure in being alive, even in the midst of a community of people required to see her as a threat to their sanctity. Babi was impossibly good-hearted. She was cheerful about absolutely everything, greeting everyone and everything with the same happy grunt. "Hah," I realized, was not just my morning call but an all-purpose salutation, something like "Howzit?" or "What's happening?" She used it on the villagers, on dogs she met on the beach, or on a coconut that fell to earth nearby. It was her affirmation of life taking place. Wherever she was, was the best of all possible worlds. She was an incurable optimist.

Even so, I couldn't help noticing that she was not averse to gilding the lily. If she found something really good, such as a perfect nut or a tasty tuber, she would trot around for hours, carrying it in her mouth, prolonging the pleasure, audibly humming with delight. I have never met anyone of any kind who got greater joy out of just being.

Later, on Sulawesi, I went to look for Babi's ancestors, discovering several stocks of Sulawesi warty pigs, all of which clearly shared her genes. The likeness was so obvious that I considered the possibility of sending her a suitable mate, but the logistics of shipping a wild boar to a Muslim village through the hands of several Islamic middlemen on boats owned by people of the same faith proved insurmountable. So I left Babi to her own formidable devices.

During the next decade, I was able to lead several voyages to the eastern islands of the Arafura Sea, and to make expeditions up the rivers of Irian Jaya and into the highlands of New Guinea. Everywhere I saw pigs that strongly resembled Babi. They were not identical, but they clearly carried a lot of her blood and shared the same sangfroid, and it gradually dawned on me that there was a pattern in all this.

FORTY THOUSAND YEARS AGO, HUMANS WERE HUNTING WILD PIGS ON THE ISLANDS. ABOUT 10,000 years ago, the Sulawesi warty pig was being domesticated and traded with other islands. At around the same time, a similar process was taking place on the continent with the Eurasian wild boar. What exactly happened after that is still confused, but it is clear that the boar *Sus scrofa* extended its habitat out on to the continental shelf, swimming or being taken to the islands by breeders and traders. There are remnants of those early movements in herds of feral boar in several places.

By about 4000 BC, these immigrant pigs had begun to mix with the locals, particularly with the warty pig *Sus celebensis,* which was also being domesticated and distributed. Once again, the evidence is patchy, but by 2000 BC it looks as though a hybrid pig was created, an animal that inherited the best features of both species, combining the dignity and courage of the wild boar with the curiosity and geniality of the warty pig. Very good nick indeed.

I am well aware of making too much of my short experience with a single sow of this breed, but it seems to me that the *scrofa/celebensis* mixture was a very potent and desirable one. It was the eastern equivalent of the Western pioneer pig, an animal as friendly as the razorback was feisty, a pig tailor-made for its place and time.

Today, Indonesia is 90 percent Muslim and pigs are not welcome everywhere, but this is a cultural artifact of the last millennium. So is the fact that pork today is eaten only in those islands that now happen to be Christian. They do so because they are not Muslim, and because they have always raised and eaten pork. Pigs are part of an earlier indigenous pagan way of life in what is left of a trading system that has existed across the archipelago for perhaps eight millennia. And Babi's strange existence as the sole survivor on one eastern island that happens now to be officially Muslim is a remnant of that loose association in which pigs of a new conformation were flowing steadily eastward from Malaysia to Melanesia.

Most people living now between the islands of Sumatra in the west and Timor in the east are more or less Malay, belonging to a people who moved there out of China in about 2500 BC. These are seafaring, slight, industrious, rice-eating, olive-colored people, with a nature that tends to be somewhat reserved. Beyond Timor, islanders are taller, darker, curly haired, and altogether more demonstrative. And in New Guinea, New Britain, the Solomons, and the other islands of Melanesia, the people are lean, black, boisterous, broad-nosed, and extravagantly outgoing.

There is a clear shift. The farther east you go, the less reserved the people become, until you encounter the Papuans, who are everything Jung had in mind when he coined the term "extrovert" for people who are of an obliging, willing, and open nature. The first time I walked into an Asmat village on the Casuarina coast and was overwhelmed by a hundred friendly hands, fifty broad grins, and instant acceptance to every home, I was reminded of Babi. There were no pigs in that particular swampland, but these loud, wide-open, amiable, accommodating, and approachable villagers were her kind of people—soul mates. This was where she belonged.

And that is precisely what I found everywhere in Melanesia where there are pigs that look like Babi. They play a vital and ancient role in the local food, ritual, and ceremonial exchange of natural energy. It would not be an exaggeration to say that theirs is a "pig culture."

IN THE BEGINNING, ANIMALS WERE DOMESTICATED FOR THEIR MEAT. THERE WAS NO CONFLICT WITH agriculture, because all Neolithic settlements were surrounded by more than enough fertile land to have pastures for grazing animals as well as cultivated crops for human consumption. But the success of the Neolithic revolution produced such a rapid increase in human population that, before long, a choice had to be made between growing food or raising livestock.

Most early empires chose to concentrate on planting and growing food, because this provides the greatest return in terms of energy. It has always been more efficient for us to eat plant food directly than to process it through an intermediate animal. And the animals already domesticated were worth more alive than dead, as providers of milk or pullers of plows. So animal flesh soon became a luxury, restricted to special occasions. And the most expensive meats came to be forbidden altogether, declared ritually unclean.

This was the fate of pigs in the semi-desert conditions of Palestine, Syria, Iraq, and Anatolia where, around 4000 BC, raising pigs became so costly that they posed a threat to the whole system. The result, in the form of a divine decree recorded in the Book of Leviticus, was bad news for pigs. In a long list of dietary laws is an injunction that forbids the eating of "any animals that are cloven-footed, but do not chew the cud." There is only one animal of that description in the entire Middle East—the pig.

This taboo was not restricted to the ancient Israelites. It was imposed in various ways, for a variety of infractions, on all the pastoral nomads of North Africa and Central Asia. There is only one condition that all these people had in common: They lived in marginal areas where pigs had become a liability.

The anthropologist Marvin Harris points out that because pork was so delicious, it had to be put beyond temptation. "Thou shalt eat pork sparingly" wouldn't have worked. Nothing less than a total taboo would do. Since that seems unnecessarily harsh to us now, it has been suggested that the prohibition may have been imposed also on medical grounds, perhaps because pork can carry parasites. That is true, but epidemiological studies show that pigs seldom transmit trichinosis to humans, even in the tropics. Cattle and sheep in warm climates carry far more serious diseases. Harris acknowledges the power and persistence of religious traditions, but he suggests that in the long term, these alone seldom prevail against starvation or radical cultural change, unless they also make ecological and economic sense.

Indeed, it would be foolish to underestimate the erosive influence of affluence. I once worked for an Austrian veterinarian who built a small zoo just outside the port city of Eilat in Israel. It was privately funded and survived only because, behind the scenes, he kept a very fertile Berkshire sow whose offspring fetched high prices among ham-hungry kibbutzim just up the Jordan Valley.

A billion of us now—that is one for every domestic pig in the world—define ourselves in part by the fact that we believe pigs to be "unclean." Mosaic and Islamic law exclude pigs in any form in the diet, on the apparently flimsy pretext that they "fail to chew the cud." There is, of course, more to this prohibition than meets the eye. It made ecological and economic sense at the time, but it also served as an important rule of separation. It was, and still is, a simple and symbolic way of maintaining a separate identity in an increasingly complex world, of demonstrating difference and holiness at times of sacred ceremony, which are almost always celebrated by a feast or some other conspicuous use of food. Pigs made easy scapegoats.

They still do. Ethiopic Coptic Christians, Mandean Gnostics, and Syrian and Canaanite cults all have prohibitions against eating pork. So do the Kitan people in Manchuria, on the ground that their founding father had a piglike head. The Scots once turned up their noses at bacon, because pigs were believed to embody the devil. In Fife, the very mention of the pig was enough to bring disaster at sea, and only cold iron was strong enough to keep such bad spirits at bay. When the story of the Gadarene Swine was read aloud in any coastal kirk, every fisherman in the congregation would feel for the nails in his boots. The boar was used in Norman sculpture to symbolize all the forces of evil ranged against Christianity. In France, the name "pig" was avoided entirely, and the animal was referred to instead as the "grunter" or "the beautiful one." And in some parts of China, pigs are still and mysteriously known only as "the long-nosed general."

In Melanesia, the constraints are very different. Kaulong people in New Britain regard anyone who refuses to eat pork as inhuman. The only excuse permitted there is one that forbids a pig owner from eating his or her own animals. This honor is extended to traditional "pork partners," people with whom pigs are traded. Pork is an important symbol of political and social power there. It defines a man's wealth and influence and gives him access to significant inter-group activities where pork partners can be established.

The number of pigs anyone can have is determined by several natural factors. The most important of these is available plant food. In the Central Highlands of New Guinea, this means yams, taro, and sweet potatoes planted in hillside gardens. The territory available to Tembaga people has a limited carrying capacity, which has to be shared. From long experience they know that a community of around 300 people can afford to support about 200 pigs. That is a comfortable ratio for a good year, but if need be, pig numbers can be allowed to fluctuate so that more are raised to meet the needs of special occasions.

Throughout New Guinea, people try to plant more crops than they need, so that there will be enough if the weather is bad. In a tropical environment, none of the crops can be stored, so any surplus is automatically fed to the pigs, a practice that amounts to banking produce in the form of pig flesh, keeping reserves on the hoof. In bad years there is no surplus and pigs have to be sacrificed, either by trading them away or by making generous gifts to distant and better-placed communities, who will one day reciprocate and return the favor. If this doesn't suffice, then the best hope is to throw a truly extravagant pig feast and celebrate your way out of trouble.

This sort of ambiguous behavior led many early anthropologists to accuse Papuans of "mismanagement," "economic inefficiency," and "ostentatious waste." More sympathetic recent studies have come to appreciate that the importance of pigs in Melanesia very often has more to do with political, spiritual, and traditional calculations than with simple profit and loss.

Pigs can be sacrificed to ancestors and ghosts, to carry out essential rites in times of crisis, or to win prestige competitions. They facilitate construction projects, the distribution of property, and the reduction of inequities. And sometimes they just feed people. Each of these practices brings its own kind of reward, and none of them are wasteful luxuries.

If nothing else, they are ecologically sound, providing a very sensitive system for regulating pig numbers. When, for instance, a pig population reaches the point of becoming a menace to people's gardens, this automatically triggers the need for a great festival, which can result in the slaughter of hundreds or even thousands of pigs and startling feats of pork consumption. These festivals are often referred to as "singsings." They involve a long series of meetings and assessments, they revive appropriate rituals and ceremonies that only take place on such occasions, and they bring people together in a way that no other purpose can.

THERE WERE FLAKE AND BLADE CULTURES ASSOCIATED WITH PIG BONES IN THE ISLANDS AT LEAST 10,000 years ago, and some clearly domestic pigs in New Guinea 5,000 years ago, but most introductions seem to have taken place in the last 3,500 years.

Prehistoric burials in Melanesia and Polynesia include pig tusk bracelets, perforated pendants, and nose pieces of pig bone. The most valued pigs have always been boars with well-developed canine teeth, and there is good archaeological evidence going back at least 3,000 years of skulls in which the upper canine teeth have been removed to allow the lower canines to grow into a full circle. This takes at least ten years, which suggests that these animals must have been kept that long under some sort of domestic control.

THE HAIRLESS BABIRUSA IS peculiar in almost every respect. It is so different from true pigs that it might be more closely related to the hippopotamus.

Big tuskers are still sacrificed at great events or "pig-outs" and their tusks turned into special, highly valued ornaments, usually worn by a man who has killed in a head-hunt. In some areas, they are worn or carried in the owner's mouth in displays that become necessary when tensions are high. The message they send then is "Danger! Watch out for me, I can bite like a boar."

Most tribes have some sort of annual ceremony that involves the killing of pigs. This usually takes place in the "good time," when the rains stop and rivers that normally isolate people become passable. As a rule, sacrifices tend to take place in the early morning after an all-night song ceremony that is called "singing with pigs," which often includes wedding exchanges, burials, and peace-making rituals.

Pigs everywhere in Melanesia are considered to be special in a way that puts them into a category that includes no other animal. Pigs alone are believed to have souls something like ours, which give them an essence that accounts for their existence as social and conscious beings. So when a man dies, one of his male pigs may be sacrificed and its skull paired up with his own so that they may travel together to the ancestors. In many areas, pigs are killed only on the burial ground, so that their blood may wash down and feed the bones of those lying there.

Today, pig-owning in most areas is still taken very seriously. It can be done by men or women, and it goes well beyond just feeding to include mandatory grooming and training. This begins with a bonding procedure that involves blowing powdered lime into a piglet's nostrils to make it forget its mother's odor. This is obviously not necessary for piglets that have been suckled at an owner's breast, a practice that is still quite widespread when sows farrow more young than they can feed. Piglets of this kind are carried about by their wet-nurses in string bags or put on a leash for long enough to train them not to go into gardens. Pigs learn very quickly, but it takes skilled husbandry, and perhaps a little magic, to educate a pig to the point where it has no fear of humans or the forest and can recognize the boundaries of both. Then it can be turned loose to roam during the day and to return at dusk.

The promise of grooming is one few pigs can turn down. In many villages, sunset is announced by high, nasal calls of "Na, na, na," "Come to me, me," which results in an avalanche of quivering pig flesh, rushing to homes and hearths, not just for dinner but for a vigorous scratch behind the ears and on all the other parts that a pig finds it hard to reach for itself.

I was amazed by how well behaved the pigs were in every community I visited in Melanesia. A good part of this must result from the extraordinary rapport that exists between island pigs and their owners, but I suspect that part of the secret lies in the fact that renegade pigs are very quickly eaten.

The enormous ritual and social significance of pigs in New Guinea invites comparison with the symbiotic relationship between some East African pastoralists and their cattle.

The Karimojong, Suk, and Maasai run large herds of cattle, but although they do take milk and tap blood from the herd, they generally prefer to keep their reserves "on the hoof" and very seldom eat their own stock. This survival strategy has social and political functions that are just as important as the dietary ones. They accumulate livestock in the same way that Papuans collect pigs and use them in a variety of transactions, such as bride payments and the exercise of traditional influence.

The pig complex and the cattle complex differ, however, in that cattle demand a great deal of attention from the men, particularly the young men, preventing them from marrying until they are in their thirties and denying them any real power until they become elders. There is nothing like this gerontocracy in Melanesia, simply because pigs are easily looked after by women, and there is open competition between younger and older Papuan men.

A second difference lies in the fact that cattle never compete with humans for food. Pigs sometimes do, and they can monopolize male attention when it becomes necessary to arrange great feasts that serve to reduce the pig population.

The third contrast has to do with milk. Cattle produce an excess of milk as a renewable resource. Pigs don't. That means that Melanesians do not get protein as often as pastoralists do, and they need to make good by killing pigs from time to time. All in all, both systems are far more complex and far less inefficient and backward than they may seem to be on the surface. New Guinea pig exchanges, in particular, have been misunderstood on this account. To the outsider, it would seem a lot more sensible for everyone to eat their own pigs. But this doesn't begin to allow for ecological differences and all the benefits that are nonutilitarian.

Paula Rubel and Abraham Rosman, in *Your Own Pigs You Must Not Eat,* cross this awkward cultural divide by quoting from John Huston's Oscar-winning screenplay for The *Treasure of the Sierra Madre:*

HUMPHREY BOGART:	Give him some tobacco. *(Walter gives the Indian his tobacco pouch. The Indian takes it and offers his own.)*
BOGART:	*Gracias.*
WALTER HUSTON:	We give them our tobacco. They give us their tobacco. I don't get it. Why doesn't everyone smoke his own?

That would take half the fun out of it.

While we are on the subject of fun, this seems an opportune place to take a brief look at the last and most unlikely wild pig of them all.

Babyrousa babyrussa is the babirusa, from the Malay *babi* for "pig" and *rusa* for "deer." This little thesaurus boils down in the end to an animal described as a "deer-pig," because its tusks curl into antlerlike structures.

The babirusa is a long-legged, hairless wild pig that is peculiar in almost every respect. Its general appearance is of a cylindrical animal, with blue-gray, deeply wrinkled skin and small eyes and ears, that looks newly born, or perhaps ev.en fossorial. It is almost hairless and, like a deer or an antelope, it lacks the direct gaze of most pigs, preferring instead to present its profile, which is wartless and bemused—reminiscent of those flat, hieroglyphic portraits of minor officials in dynastic times.

But the most conspicuous feature of the babirusa is two pairs of canine teeth that curl back toward the forehead in great sweeping curves to form a sort of face mask. The lower tusks are simply enlarged, but the upper ones are bizarre. Their sockets have rotated so that the teeth grow through the top of the muzzle to emerge from the middle of the animal's face and curve back in an overreaching arch. The effect is startling and the function is obscure. This structure is unique in the animal world and so brittle that it is useless for combat or defense. Local

legend insists that babirusas hang by these tusks from the branch of a tree when they need a safe place to sleep.

There is no real consensus about babirusa classification. Their oddities include a stomach more like that of an ungulate, except that they don't chew a cud.

THE UPPER PAIR OF tusks in the babirusa are bizarre, growing through the roof of the snout in twin arches whose function remains obscure

They are too thin to be typical pigs and they have some anatomical features that suggest they might be more closely related to hippos. Babirusas occupy a subfamily of their own at the moment, and there is general agreement that they have been following their own evolutionary line for at least 25 million years.

Babirusas are endemic to the rainforests of northern Sulawesi and seem to have been isolated there by deep, fast-flowing straits ever since the continental shelf was flooded.

These enigmatic animals grow to be about 3 feet long and 3 feet tall, weighing up to 200 pounds. Females are a third smaller. They are omnivorous but seem to concentrate on fruits and nuts, showing a strong preference for mangoes. They sometimes catch and eat small mammals. Among their anomalies is the lack of a rostral bone at the tip of the snout, but they nevertheless indulge in rooting behavior in soft soil. Males do this vigorously in new or disputed territory, kneeling down and sliding forward on their chests, plowing long furrows, perhaps leaving scent trails from tusk and cheek glands.

Babirusas appear to be diurnal and to live in small family groups. Their gestation period is about 150 days, which is long for pigs, and their litter size is very small. They have only one pair of teats. Babirusas build nests among the roots of canopy trees or in convenient caves, and they run fast and swim well.

Little is known about their behavior in the wild, but in captivity they seem to be contentious. The males make sudden, unprovoked charges at each other, though they seldom come into contact. These encounters usually end in formal circling maneuvers with noses held high, but if they do touch, one lays his snout and chest on the other's forehead, paddling until the other submits with a high-pitched squeal. In a very small number of aggressive bouts, the combatants stand up on their hind legs and box each other with their hoofs in a deerlike conflict that is more or less ritualized and unique among pigs.

Babirusas are known to have lived to be twenty-four years old. A herd book is being kept by Berlin Zoo that records all animals in captivity. The species is rare and endangered in the wild but has been declared a "flagship species" for conservation purposes in Indonesia's national parks.

This extraordinary animal, which it is hard to think of as a pig, has been hunted on Sulawesi for thousands of years, mainly as a source of meat. It does not breed fast enough to be useful in domesticity, and it cannot be crossed with true pigs, because its chromosomes have at least five anomalies with no equivalent in other pig species. Babirusa survive, it seems, only as fascinating curiosities.

THE BIBLE OF PIG CONSERVATION IS A 1993 WORLD CONSERVATION UNION (IUCN) STATUS SURVEY and Conservation Plan titled *Pigs, Peccaries and Hippos,* which encompasses all the living members of the suborder Suiformes. In this survey, put together by thirty-six specialists, is a useful outline of each species and its prospects.

The babirusa is indeed peripheral, confined largely to one peninsula on northern Sulawesi and a few small islands supporting a combined population of less than a thousand individuals. It seems to have played no part in pig evolution, other than its own lonely and idiosyncratic passage as a weird antelopine sideshow to the history of the pig family.

The bearded pig is another curiosity, obviously related to the warty pigs but alone in its extraordinary lemminglike migrations. These mass movements have left it scattered across the Malay Peninsula, Java, Borneo, and several smaller islands, where it continues to hold its own in the last forested areas. For the moment, it is relatively abundant in all its main haunts and must number somewhere in the high five figures.

The other islanders—Javan, Visayan, and Philippine warty pigs—are marginal. They continue to hold out against deforestation in most of their traditional areas, taking refuge in high, remote, or protected patches, but in all of these they are endangered by burgeoning human populations. Numbers are unknown but are perhaps in the high four figures. The only encouraging prospect is that of *Sus celebensis,* whose cheerful line continues in the form of semi-domesticated and feral populations in Melanesia.

In Asia, the pigmy hog ekes out a precarious existence on a few relict grasslands in the southern foothills of the Himalayas. There cannot be more than a few hundred of these tiny relatives of the wild boar, building their last, optimistic nests in Assam, guarding a genetic resource that could be lost in just one or two brush fires.

The Eurasian wild boar itself is doing well, perhaps even increasing in numbers, repopulating old habitats by learning how to live with humans without conflict. Even if they were to vanish from the entire continent, there are probably enough resourceful feral boars in the rest of the world to recreate and repopulate their home areas again. Their numbers are in the high six figures.

The African pigs are not doing too badly. The common warthog is abundant in the parks and reserves of South and East Africa, holding its tail flag high, trotting and grazing its way across the savannah. Other stocks in West Africa are less secure, but their relative, the desert or cape warthog, is now extinct in the Cape and greatly endangered in the desert Horn of Africa, where it does not exist in any protected area.

The less approachable bushpig and red river hog are nevertheless locally abundant. Their nocturnal habits keep them out of harm's way, even in well-populated areas, though they are subject to trapping and hunting with dogs. Both species are so adaptable that even deforestation doesn't seem to reduce their numbers. They may benefit from the opening up of forests to provide a wider variety of secondary habitats. Both bushpigs flourish when offered access to cultivated crops.

The larger and equally elusive forest hog is more dependent on gallery forest and exists now only in pockets of high-altitude primary growth. These areas are not yet densely populated, but logging ventures and new roads are making access easier for hunters and poachers. These giant pigs are not yet threatened but they need more protected parks if they are to survive.

And as far as peccaries are concerned, these old-fashioned, New World suids have one widespread and successful species in the collared peccary, one reasonably well-protected rain-forest species in the white-lipped peccary, and one recently discovered and even more recently endangered primitive species in the giant peccary of dry-thorn forests in Argentina.

Taking all sixteen species of living wild pigs together, their total global population may lie somewhere around 10 million animals. There are more living things in a school of herring or a handful of soil, but pigs have had a disproportionately great influence on our lives.

There are pig bones bearing the marks of butchering in the camps of ancestral *Homo erectus* made more than a million years ago. Our own species Homo sapiens was busy hunting and eating wild pig in rock shelters along the shores of southern Africa 125,000 years ago. And someone enjoyed roasting bearded pigs in caves on Sarawak 40,000 years ago.

Wild pigs have been the most commonly eaten wild animals everywhere, and they may even have been the first animals we chose to domesticate. Because pigs are so versatile, adaptable, and fertile, they quickly became the most abundant and widespread hoof stock in the world. As a result, the diversity of wild and domestic, feral, native, and introduced pigs has produced patterns of distribution and interrelation that are awesomely complex.

There are now 1 billion domestic pigs in the world. Of these, 60 percent are in Asia, 20 percent in Europe, 15 percent in the Americas, and the rest in Africa and the Pacific islands. And in some of these areas, there are still functioning "pig cultures," whose lives and traditions are entirely governed by the pigs with which they share their homes and gardens.

Pigs, wild and domestic, like it or not, are a force in our lives. They are everything, everywhere, and ever ambiguous—massive or dainty, finicky or fat, stolid or effervescent, but never anything less than compelling. They are animals for all seasons, found everywhere except the continent of Antarctica. And wherever pigs may be found, the one thing that everyone agrees on is that pigs are far smarter than any cloven-hoofed ungulate has a right to be.

IN EIGHTEENTH-CENTURY LONDON, FOR INSTANCE, LEARNED PIGS WERE VERY MUCH IN VOGUE. IN 1785 Sadlers Wells introduced a "sagacious pig who reckons the number of people present, tells by evoking on a Gentleman's watch what is the hour and minutes, and distinguishes all sorts of colours." In 1789 Joseph Strutt reported on another smart pig, "which even at the polite end of town, gave great satisfaction to all who saw him, and filled his Tormentor's pocket with money." He arranged words by picking up letters written on pieces of card in reply to questions posed to him, "providing another proof of what may be accomplished by assiduity and perseverance."

Even Dr. Johnson was amused. "Pigs," he said, "are a race unjustly calumniated. Pig has, it seems, not been wanting to man, but man to pig. We do not allow time for his education; we kill him at a year old." On learning that the town's latest pig prodigy was three years old, he concluded with satisfaction, "There, the pig has no cause to complain; he would have been killed the first year if he had not been educated."

But the best known of all educated pigs was Toby, who amused audiences with his ability to do arithmetic, spell words, and compose poems in honor of Roger Bacon. The early romantic poet William Cowper saw the pig performing at Norwich and complained, "I have a competitor for fame, not less formidable, in the Learned Pig. Alas! What is a tutor's popularity worth, in a world that can suffer a pig to eclipse his brightest glories?"

Another poet of the time, Thomas Wood, chose to compete with this pig directly in a long poem of his own called "The Lament of Toby," which includes the lines:

> *Oh, why are pigs made scholars of*
> *   It baffles my discerning,*
> *What griskins, fry and chitterlings*
> *   Can have to do with learning.*

Toby performed in public with a frilled collar around his neck, and it was not long before he got involved in show business instead of education. So it was inevitable that he should write an auto-biography entitled *The Life and Adventures of Toby the Sapient Pig, with his Opinions on Men and Manners*. It was embellished with an elegant frontispiece, showing the author in his literary sty, and published by his teacher Nicholas Hoare, who is acknowledged in the text as "my preceptor, whose indescribable solicitude and promptness of application were never equalled by any of the learned teachers at either of our universities."

In the nineteenth century, Jemmy Hirst, a Yorkshire eccentric, rode a bull out hunting and used Hampshire pigs as pointers, while a farmer who lived near Saint Albans went to market at a brisk trot in a cart pulled by four large hogs. The circus, naturally, was not far behind. In the fifteenth century, Louis XI of France, whenever he felt gloomy, chose to be cheered up by a troupe of pigs festooned with pants and ribbons, who danced for him to the accompaniment of bagpipe music. Lord George Sanger at the turn of the nineteenth century augmented the lions, llamas, and clowns in his circus with a performing pig called Bill. Bill did all the usual hoops and jumps and arithmetic with a frill round his neck, but his grand finale was a great feat in which he balanced with just one trotter on the nose of a sea lion.

Well-schooled pigs were among the first circus performers in America as well. In the 1840s a showman called Dan Rice took his trained pig out on the road, and in the 1920s the jazz singer Josephine Baker seldom appeared without a stage full of perfumed pigs whose chore-ography had to be seen to be believed. Fred Kerslake toured every country fair with a vaudeville act that included a talented five-pig group that performed on see-saws, ladders, and carts in an act so polished that, as the *Roanoke Times* recorded, "The spectators were given the impression the pigs themselves had a well-developed sense of humour."

Indeed, they do. Pigs are natural entertainers. They don't have to be forced to perform, and they have a remarkable range of expression and breadth of talent. Hog expert William Hedgepeth allows that all pigs have a fine lyric sense and a general love of music. He quotes a circus hand who says, "The average hog can be as wise as an owl, stubborn as a mule, sly as a fox, quick as a wink, smart as a whip and, alas, as poor as a church mouse." Pigs seem also to have an uncanny sense of timing and can play the straight man in a routine, but they are not above "hogging the limelight."

PIGS LEARN VERY QUICKLY AND need no rehearsal or persuasion. They are born hams.

All this is true, but the two things that interest me most in this catalog of zoological follies are that pigs learn very quickly, needing little or no rehearsal or persuasion, and that they will spontaneously perform without tuition or a human audience. Every pig I ever met was a ham, and great company.

The lyric poet Robert Herrick, he of "Gather Ye Rosebuds While Ye May," kept a clerical pig that followed him everywhere and may have contributed to his ejection from a Devonshire vicarage by the Puritans. Lord Gardenstone, an eminent Scots lawyer, kept the best-known legal pig, which shared his bed.

Captain Basil Hall of the Royal Navy kept a white sow as his ship's mascot, entitling her to the usual daily ration of grog and a proper naval burial at sea. But the most aristocratic pig ever was perhaps Cupid, the favorite companion of the earl of Edgecumbe and his countess, who took her with them to meals and introduced her to society in London. Cupid was buried in the end in a gold casket beneath an opulent monument bearing an epitaph composed for her by the poet-physician John Wolcot.

HISTRIONICS ARE NOT A RELIABLE SIGN OF INTELLIGENCE, BUT ANY ACT THAT CAN IMPRESS A down-to-earth farmer needs to be taken seriously. In his classic book of 1789, Gilbert White—clergyman, farmer, naturalist—tells of a sow that his Hampshire neighbor described as sagacious and artful: "When she found occasion to converse with a boar, she used to open all the intervening gates, and march, by herself, up to a distant farm where one was kept; and when her purpose was served would return home by the same means." The gentleman-farmer Sir Walter Gilbey reported that he once saw "an intelligent sow pig about twelve months old, running in an orchard, go to a young apple tree and, shaking it, pricking her ears at the same time as if to listen to hear the apples fall. She then picked the apples up and ate them. After they were all down she shook the tree again and listened, but as there were no more to fall, she went away." Yet another farmer, the author of a book about pigs, dropped his bucolic guard far enough to admit that "pigs have a keen sense of the absurd and will suddenly take off in a collective giddy fit, twirling round and round to the accompaniment of hoarse pantings, guffaws, it might almost be said, of merriment."

Such playful behavior by pigs is epidemic. It smacks of a definite lack of constraint and the deliberate pursuit of pleasure instead of dreary routine. Pigs enjoy novelty enormously and will go a long way to find it. Every now and then, some pigs get so desperate for variety that they make much-publicized breaks for freedom.

Charles Dickens, amazed by the number of pigs running loose in the streets of New York in the mid-nineteenth century, wrote in admiration of a particular boar he encountered on Broadway: "He leads a roving, gentlemanly, vagabond kind of life, somewhat answering that of our club-men at home. A free-and-easy, careless, indifferent kind of pig, having a very large acquaintance among other pigs of the same character. In this respect a republican pig, going where he pleases and mingling with the best society."

Nothing changes. The *New York Times* recently reported the sight of two adult hogs trotting alone down the streets of Manhattan in the early hours, and it published a later report of the same two gentlemen of leisure strolling nonchalantly along the foreshore of Saint George on Staten Island. William Hedgepeth wonders: "How in the world did they do that? Hogs can swim well enough, but that would have been stupid. Clearly they got aboard the Staten Island Ferry, but how? And where did they get the fare? Merely contemplating such a thing can lead to madness."

The chase goes on. In 2002, a 250-pound large black boar called McQueen contrived his great escape from captivity by leaping over a 6-foot wall surrounding an abattoir in central Scotland, where he was awaiting slaughter. Police warned the public not to approach the runaway and chased him for two days until he went to ground in deep forest near Dunblane. The story of his bid for freedom led in the end to a fund large enough to ensure McQueen a comfortable future on a friendly local farm.

But the most poignant of all pig escapees has to be a young one known only as Pig 311, who jumped ship in the Marshall Islands in 1946. She succeeded in swimming several miles to shore and was eventually rescued and spent the rest of her life in the National Zoological Park in Washington. Pig 311 grew to be a 600-pound sow and was normal in every way except for the fact that she was sterile—probably because she chose to make her break and land on Bikini Atoll at the precise moment an American atomic bomb was tested there.

Anecdotal evidence, and there is plenty of it, points to the conclusion that pigs seem to be at least smart enough to have attracted centuries of interest and attention. Their evident playfulness, sociability, and curiosity make them unusual among hoofed animals. Pigs stand out in any farmyard as animals with a mind of their own.

But the accounts of learned and sagacious pigs do not necessarily prove high intelligence. It is more than likely that performing pigs such as Toby are the porcine equivalents of Clever Hans, a horse that seemed to have extraordinary talents until he was discovered to be picking up very subtle cues from his trainer. If the man didn't know the answer to a question, then the horse wasn't able to provide it either. Such minute attention to unconscious signals is obviously within the capability of pigs, whose communication with one another depends on similar minutiae.

We need better evidence, properly controlled and far more carefully designed, but the appropriate literature on pig intelligence is strangely lacking. There seems to be a presumption that pairing the words "pig" and "intelligence" constitutes a contradiction, a scientific oxymoron. The best that science seems willing to allow at the moment is "pig cognition," which plays safe by considering perception instead of knowledge, sensation instead of intellect. I find that unnecessarily timid, but it is at least a place to begin looking for what it is that makes pigs so interesting.

The benchmark for me is a paragraph written by the agriculturist William Youatt in 1855:

> In general, there is nothing in the life of
> a hog, in his domesticated state at least, which
> calls for any exercise of reasoning powers.
> All his wants are anticipated, and his world is
> limited to the precincts of his sty. Yet even
> in this state of luxurious ease, individuals have
> shown extraordinary intelligence. There are
> anecdotes enough to prove them possessed
> of memory, attachment, and social qualities;
> but at present the system of treatment affords
> no scope for the development of any but
> mere brute instincts.

Just so. But it wasn't until 1914 that anyone tried to pin down this elusive pig quality. The pioneer in question was Robert Yerkes, one of the principal developers of clinical psychology. He had the courage to study intelligence at all levels of life, from jellyfish to great apes, and is best known for his work on chimps at the Yale Laboratory for Primate Biology, where he studied the natural bases of behavior. But twenty years before he started on that, he was already experimenting with pigs.

Yerkes designed a multiple-choice system in which different species could be challenged with the same task. This involved finding a food reward hidden in a line of identical stalls set up in a row. The object was to find the right stall in the least number of trials. The test was repeated 40 times, and it went on each time until the reward was claimed.

All that was simple enough, but there were complications. The reward was put in a different stall at the end of each set of ten tests, moving say from "the first stall on the right" to "the second stall on the left." And one or more stalls were locked and taken out of play in a pattern that changed with every test, so that subjects had to make a mental shift from "the second stall on the left" to "the second *available* stall on the left." Yerkes also included a punishment for every mistake, which involved being locked into the empty stall for a minute after each wrong guess.

Taken altogether, this fiendishly baroque experimental procedure was designed to test the patience of a human chess master, but it was at least the same for every subject of every species. Yerkes tested it first on rats, then on ring-doves and crows, looking always for evidence of an ability to find and memorize patterns, benefit from them, and make appropriate adjustments each time the rules changed. What he wanted was proof of an ability to have and to hold an abstract thought, an "idea."

Rats did reasonably well. Ring-doves were patient but slow to adapt. Crows were good, almost intuitive, but got bored rather quickly. Then Yerkes tried pigs, Chester white pigs in fact, in a larger version of the apparatus set up outdoors in an orchard at Harvard University's field station. "The pigs," said Yerkes, "proved far more satisfactory subjects than we had dared to hope. Indeed, they worked so steadily and uniformly through the investigation that there was practically no loss of time. It is chiefly because of this unexpectedly favorable relation of subject to method, that we were enabled to obtain such numerous results."

The results were rich and rewarding. In Yerkes's own words again: "Our results indicate for the pig an approach to free ideas which we had not anticipated. While hesitating to claim that we have demonstrated the presence of ideas, we are convinced that the pig closely approaches, if it does not actually attain, to simple ideational behaviour." Which is cautious psycho-speak for "The pigs got it!" And they did so faster than several human students who were later given the chance to meet the same challenge.

As far as I can discover, Yerkes never worked with pigs again, and neither did his students. The rest of the psychological behaviorists went on to increasingly boring and repetitive studies on white rats, which had little relevance to the problems of survival in the real world, and pigs were left floating in a scientific limbo.

It is hard to understand why. That early porcine performance begged for confirmation and extension, but nobody picked up on the titillating "And so . . . ," which was left just hanging there.

Serious studies of comparative behavior seem to have wandered off into a psychological ghetto presided over by Ivan Pavlov and B. F. Skinner, who taught pigeons to play table tennis and turned the subject of "learning" into something that took place by reinforcement and reward in a Skinner Box. This work was dominated, right from the start, by the assumption that the "laws of learning" are the same for all animals, and that the wide differences in brain structure have a purely quantitative significance. Skinner himself said: "It doesn't matter which animal is studied."

As a result, two-thirds of all studies on animal learning in the first half of the twentieth century were done on laboratory rats, with not a pig in sight, and this unbiological madness prevailed until Konrad Lorenz and Niko Tinbergen went back to the wild and reintroduced natural selection to the study of animal behavior.

The second half of the twentieth century saw a revival of natural history in the far more orderly and constructive framework of ethology, which not only encouraged careful observation of behavior but also called for speculation about its function, survival value, and evolutionary significance. This resulted in the construction and publication of ethograms and monographs of a wide range of animals, including a few pigs. Notable among these are an IUCN publication *The Behaviour of Ungulates,* edited by Valerius Geist and Fritz Walther in 1974, and the proceedings of a workshop on wild pigs published in a special edition of the journal *Bongo* in 1991, as a tribute to the German zoologist Hans Frädrich.

These and other publications have faithfully documented what is now known on all the living wild pigs and peccaries of the world, and they stand in direct contrast to latter-day psychological studies on pigs, which are, almost without exception, concerned only with domestic pigs bred for and tailored to laboratory studies. There is, however, little in either discipline that sheds direct light on the subject of pig intelligence until the last decade of the second millennium.

The postmodern attitude to pig intelligence has grown out of a fertile relationship between ethology and philosophy that began with Peter Singer's work *Animal Liberation,* in 1975, and the first of Donald Griffin's essays, *The Question of Animal Awareness,* in 1976.

Both books arose from ethical concern for animals and the evolutionary continuity of mental experience, and both were strongly criticized as "anthropomorphic musings without scientific merit." It is indeed difficult to demonstrate true awareness or consciousness in other species, but it is becoming more and more difficult to deny the possibility of animals having minds and using mental experience in the practice of their behavior.

The most determined effort to break this impasse has come in the last decade from a small group of life scientists with experience in ethology, psychology, and philosophy who are working hard to establish the new field of "animal cognition" and give it scientific respectability. Happily, after almost a century of neglect, pigs are once again being called in to provide crucial testimony.

One of these brave scholars defines cognitive ethology as "the synthesis of sensation and memory which produces appropriate responses." Another points out that anthropomorphism is not necessarily sinful or unscientific, because it is only incorrect if other species truly do lack humanlike mental abilities. That is certainly not proven for pigs.

The problem is very neatly illustrated by an anecdote about pig behavior. This concerns a group of young pigs who discovered they could cross an electrified fence as long as they ran fast enough together through it. Each time they decided to do so, they gathered some distance back and ran rapidly toward the fence, squealing loudly long before they reached it and continuing the cry until they could gather safely on the other side. Common sense suggests that the intrepid sounder knew that this action was going to hurt and were prepared for the pain long before they hit the fence. But this interpretation is totally rejected by science, on the grounds that it is anecdotal and lacks any experimental control and because we cannot possibly know what it is like for a pig to experience an electric shock.

That, unfortunately, is how things stand. But the barrier of willful incomprehension is under siege and cannot be kept intact for much longer. The anecdote about the fence makes me think that, at the very least, those pigs were exercising a very nice judgment in deciding that

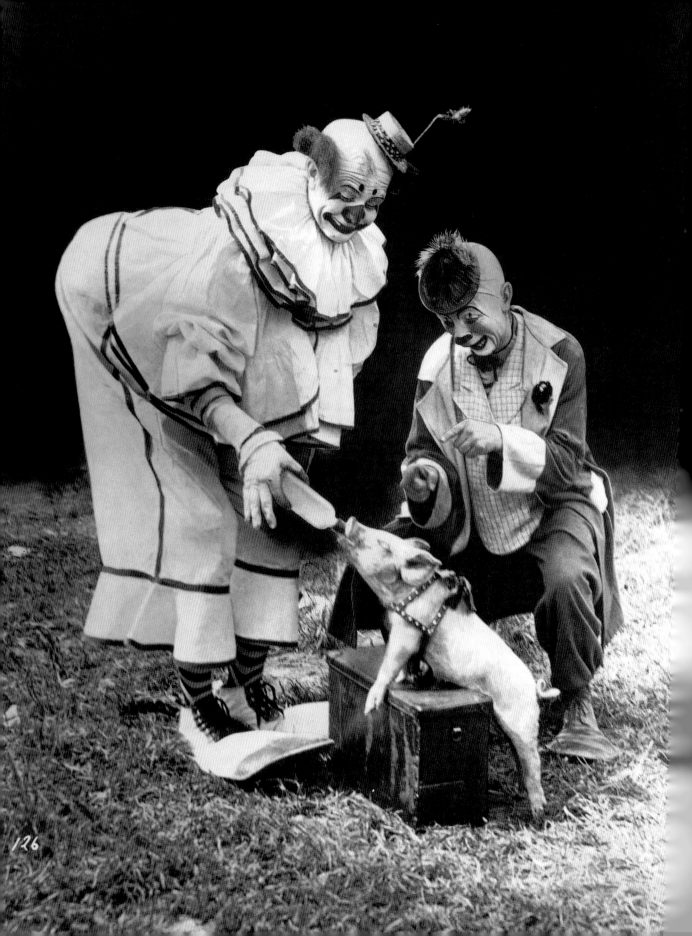

126

the pain involved was outweighed by the benefits of freedom in greener pastures on the other side. That is exactly how pigs seem to come to most conclusions.

My experience of pigs is that they do demonstrate considerable variation in behavioral strategies when faced with new challenges. And when confronted with such problems, they are flexible enough to modify their own behavior. They learn about the world as they go along.

There are natural barriers to how much pigs can develop and change, and the most obvious of these is the fact that their evolutionary history denies them a hand with which they can manipulate their environment. But there is one interesting line of research that might help us to work around this limitation.

Candace Croney and her colleagues at the Oregon State University have been experimenting with an apparatus based on a video-game task that pigs can access by means of a joystick that they move with their mouths in order to control a cursor on a computer screen. This has proved successful in allowing pigs to be tested on simple learning and memory abilities, but it could just as easily be dedicated to discovering exactly what pigs can do when liberated from their anatomical constraints. An imaginative series of inquiries using this technique could be exactly what we need to make a quantum leap into the study of the pig's mind.

BRING ON THE CLOWNS, IF YOU dare. Pigs have a habit of upstaging all human performers.

On the frontier of research into pig cognition at the moment is the ethologist Michael Mendl, who operates a pig unit at the University of Bristol. Starting in 1997, he and his colleagues looked at the foraging behavior of large white pigs in semi-natural surroundings and discovered that they have well-developed spatial memories for places where food can be found. That is true of squirrels and dogs as well, but he has gone on to show that pigs often gain extra advantage by exploiting or concealing such knowledge, and that they will on occasion employ such deceptive tactics to increase their own success.

The protocol is elegant. It involves two pigs, one of which goes into the test area and, by trial and error, discovers where food has been hidden. Later, this "informed pig" is reintroduced to the same area in company with a larger "uninformed" companion. As expected, the pig with prior knowledge found and ate most of the food. But when the experiment was repeated, the large pig simply followed the small know-all to the hidden food and commandeered it in a classic case of exploitation. That, however, was not the end of the story. Before long, the little pig got smart and waited until the exploitive bully was otherwise distracted before taking advantage of his inside information.

"There are two possible explanations," explains Mendl. "Maybe the informed pig is working out what the other pig is thinking. If so, that's a very high-level explanation. Or maybe the informed pig simply learns that if it goes directly to the food when the other pig is watching, it gets displaced."

If the first explanation is true, this means that pigs have what psychologists call "a theory of mind," something equivalent to the surprise a seven-year-old child experiences when it realizes, for the first time, that other people think thoughts unlike its own. But even if the subordinate pig exercises only a cautious delay in going directly to the food, it is using its head in a creative way. Both possibilities suggest mental activity and a willingness to adjust behavior. And, in the end, it is the smarter of the two pigs, not the strongest, that gets to put its nose in the trough.

The capacity to increase your own foraging efficiency by exploiting the knowledge of another is obviously one that would be very useful in the wild. But the combination of knowledge and wisdom is an even more potent survival strategy, and the fact that pigs kept in domesticity for thousands of generations can still retrieve such a skill the first time it is required is fascinating. The next experiment in Bristol will involve a test to find out if a pig who cannot see where food is being hidden will follow another pig that is able to see where the hoard is being concealed. The test may show that pigs are capable of understanding what other pigs know, which is far more complex and revealing.

I am afraid, however, that even this might not be enough to convince the skeptics, who insist that nonhuman animals are incapable of conscious thought or even of feeling pain. My pet warthog once stood on an acacia thorn that broke off between the cusps of his hoof, and he cried out loud enough to bring me running. There was no question in my mind about what had happened. Hoover was conscious of feeling pain, until I removed the thorn. Science, in its rush to objectivity, denies such common sense.

I believe there is already enough evidence to show that pigs do consciously process thought and are capable of abstract reasoning, but I concede that there is still not enough documentary evidence to make such a conclusion scientifically acceptable. What is needed are clearer definitions of consciousness and an experimental protocol that can demonstrate that most higher animals do exhibit this property in meaningful and different ways, according to their needs.

I suspect that what is required for us to arrive at such an understanding is not a complex new piece of apparatus or more statistics, but perhaps just a new and better way of asking leading questions. Who dreamed of elephant infrasound until Katy Payne posed her simple query, "Can elephants hear sounds below the human threshold?" Who knew anything of navigational skills among insects until Karl von Frisch asked, "Why do honey bees dance?"

What we are going to find out in the end, I think, is that all animals have the tools required to deal with the problems posed by natural selection. They would be extinct without them. And in the pig we have a true oddity, an animal whose most useful tool is its brain.

THE FIRST THING YOU NOTICE ABOUT PIGS, WILD OR DOMESTIC, IS THEIR HEADS. IN ALMOST ALL species, the head is clearly pronounced and set off against the rest of the body. In some, it represents almost a third of the entire length of the animal. And this contrast is magnified by tufts, bristles, warts, beards, tusks, and a variety of growths about the head that draw attention to it immediately. There is also a clear succession in evolution of such differentiation from the relatively undifferentiated heads of wild boar, through the conspicuous bulges of the warty pigs, to the elaborate architecture of the warthog's head and snout.

In some cases, the head is almost theatrical, emphasized by colors or patterns that make it look like a medieval mask, something worn in battle or a religious ceremony. In fact

it serves precisely that function, exaggerating for the purpose of visual display, threatening rivals in ways that make male pigs seem larger and more fearsome than they actually are.

PIGS ARE TRUE ODDITIES IN THE animal world. Their most useful tools are their brains.

This is called *cephalization*—an elaboration of the head, both externally and internally. Beneath the skin, the cranium is sculpted into a wonderful bony helmet that ranges from smooth armor plating to ornamental knobs and bosses, all designed to protect the brain within. Pigs have a proportionately larger brain than cattle, sheep, or antelope, and they also possess more of that part of the brain that governs an ability to reason. Pig brains are not unlike our own. There is little but size to tell them apart, and the causes and consequences of our grossly inflated brains are still far from obvious. That's one of the great mysteries about cephalization.

Separate studies of domestic animals at Cornell University in New York State and the Jackson Laboratory at Bar Harbor in Maine have shown that in maze tests dogs score well, chickens and horses manage more slowly, sheep mill around and never finish, and cats refuse to participate altogether. But pigs come out on top every time, because of "independent thinking and an ability to figure out a problem for themselves." And these were domestic pigs, with smaller, less exercised brains than wild pigs. Farmyard pigs, despite their great body size, have a brain that is about 20 percent smaller than that of their wild counterparts.

The difference between them is dramatic, but the change is reversible. After just one generation of feral life, you can see differences in the skull as the brain begins to inflate again. It is not difficult to distinguish between animals reared in "pig parlors" and pigs of the same breed reared outdoors and allowed to forage for themselves and lead a more or less natural social life.

There are ways of counteracting the stultifying effects of domestication. In Sarah Boysen's pig unit at Ohio State University, they seem to be going well beyond the usual academic restraints on most studies on learning. Pigs there are performing complex tasks in response to gestural and vocal symbols that lead to the storage of abstract information in patterns that come very close to the use of language. And they look different.

This doesn't surprise me. Pigs are extraordinarily plastic and have a happy capacity to take things in their stride, even after long domestication. It looks as though the genes responsible for wild characteristics and functions are still there, even after hundreds of generations, bottled up somewhere in the genome, biding their time, just waiting for the stimuli necessary to turn them on again. There is a wild boar lurking in the body of every cottage pig.

Over the past few years, laboratory experiments have provided convincing evidence that some organisms really can save up genes, or mutations in genes, for rainy days. Both plants and animals appear to be able to control evolution in the sense that they accelerate or decelerate variation as required. The trigger to this marvelous mechanism is a protein called hsp90. It is a "chaperon" substance, one that binds to other less stable proteins and tidies them up, out of the way, putting them on hold and keeping them quiet until the time is right.

This chaperon was discovered in the geneticist's best friend, the fruit fly. At the Whitehead Institute for Biomedical Research in Cambridge, Massachusetts, they noticed that something was producing "cryptic" genetic effects in a few of their flies. Bizarre changes were taking place, but they only occurred when hsp90 was absent. When the protein was present, papering over flaws in the genome, everything was stable. Without it, the hidden mutations were suddenly unmasked and produced abrupt changes in shape and form. It is now thought that hsp90 may be doing the same thing in plants such as thale cress, in some African frogs, and even in bacteria such as *E. coli*.

The upshot is that, for the first time, we seem to have a mechanism for delivering variation. If hsp90 can buffer mutations that are not required, then it can also reveal these when they are necessary and useful—variations on demand that are a direct response to environmental pressures.

This also provides a mechanism for the sudden leaps in evolution that Stephen Jay Gould called "punctuated equilibrium." These are changes that involve not just one mutation but whole suites of variations that may produce sudden and dramatic alterations, leading to a whole new design.

This possibility is very new. Evolutionary biologists are still examining the consequences and implications of the idea, wondering perhaps if flaws in hsp90 may also be involved in genetic diseases such as cancer, or the loss of immunity. Research goes on, but in the meantime it appears that pigs are busy, in their own inimitable fashion, putting it all into practice, uncloaking the wild boar as occasion demands.

That is something to think about. Never mind a little gentle learning; I can imagine pigs that are allowed to hurt, cry, feel, communicate, solve complex social problems, and think all such things through—pigs with their own cultures.

Why not? Defining "culture" as uniquely human is just lazy thinking, the next logical step beyond self-awareness. Culture is an adaptive function. If California sea otters can develop a tradition of using stone anvils to open clam shells, and Japanese macaques are already beginning to carry pebbles about, looking for some appropriate use for these proto-tools, I predict that pigs are going to be found to be cultured animals in ways best suited to animals without grasping hands.

It is difficult to wrap our matter-of-fact minds around such abstraction, but the news that red river hogs in Uganda are already practicing agriculture with nothing more than their snouts should provide pause for thought. Pigs may be capable of doing almost anything. Expect the unexpected.

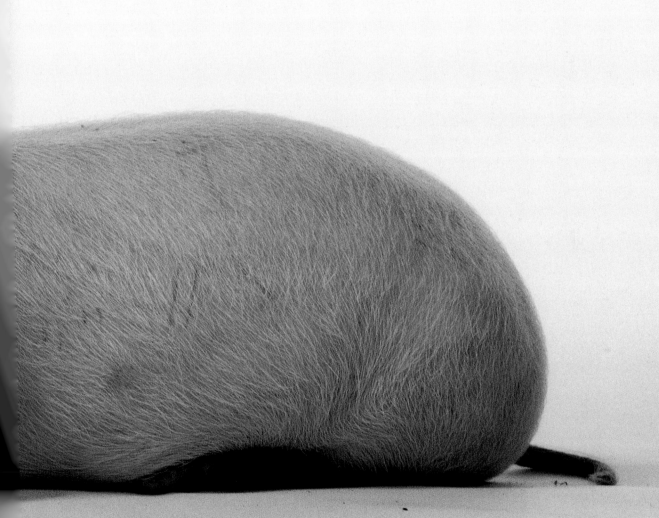

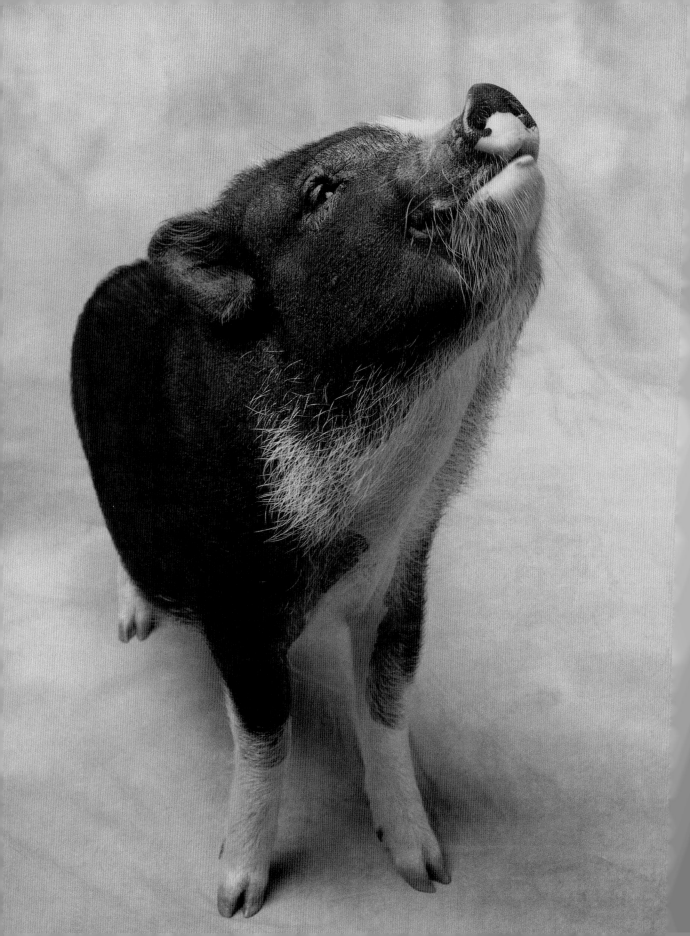

When Pigs Fly

Ethology in the last half century has fostered a wide range of studies on animal behavior in the wild. To start with, these were evenly distributed across species and habitats, as each new student looked for a subject of their own.

With time, however, some species in some circumstances proved worthy of deeper investigation and groups began to accumulate around convenient centers such as the Serengeti or the Barro Colorado, focusing on carnivores, primates, or ants, and questions began to be asked about a handful of well-studied species such as the black-headed gull, the three-spined stickleback, and the chimpanzee. Are rich results somehow intrinsic to these showcase species, or are the discoveries about them just a reflection of the depth of research being done?

       I suspect a lot of the latter, and I would suggest that there are similar species lying in wait for anyone who spends enough time with, or asks the right questions of, almost any living thing. We are still only scratching the surface of biological possibility, and we are ignoring subjects that seem to be unlikely, or just unappealing, to our disadvantage.

       Pigs have suffered from such neglect, I think. With the exception of warthogs, which are very obvious and obliging out on the savannah, wild pigs are not easy to work with, though they are certainly no more difficult or dangerous to study than dolphins or mountain gorillas.

I am at a loss to explain why pigs have had such a bad scientific press. It can't be for lack of interest or public appeal. Pig paraphernalia abound. They may even outsell bears and horses in the market for ceramic keepsakes, bumper stickers, and fridge magnets, but despite Miss Piggy's best endeavors, there are still no "teddy-pigs" to take to bed or "rocking-pigs" to ride on. There is a huge pig constituency out there. Harry S. Truman said: "No man should be allowed to be President who does not understand pigs, or hasn't been around a manure pile." And Michael Ramsay, archbishop of Canterbury, received more enthusiastic letters in a single week than he did during the rest of his primacy, when he made passing mention on the radio of his soft spot for pigs.

The relative lack of scientific enthusiasm is strange. There are more biologists studying the Japanese macaque in the field than there are on the track of all the world's wild pigs combined. Perhaps funding is a problem, but that hasn't stopped hundreds of researchers from gathering around species as lacking in appeal and personality as the black Norwegian rat. The answer must lie in the fascination:repulsion ratio that impedes pigs as a whole. They suffer, not so much from their own shortcomings as from our misguided perceptions of them, and they seldom get to show us what they are really like, because we hardly ever meet the same pig twice, which is a shame.

I have been lucky. I have had the great good fortune to see most of the world's wild pigs on their home ground and to have had a close personal relationship with three pigs on three different continents. None of these were academic, formal associations. They were pleasures that took place while I was busy doing something else. But they continue to animate my life in interesting ways. I am enriched by what amounts to a social synapse between suids and myself, an awareness that has sensitized me to pigs and piggish things and continues to be most rewarding.

For what it is worth, I believe that pigs have a message for us that is there for the asking, because we share so much already. We are both products of an omnivorous upbringing, curious, dexterous, and willing to explore new things. And, as a direct result of such open-minded, open-mouthed enthusiasm, we are what we have eaten. We are the consequences of parallel adaptation, genetically modified by long association with a wide range of plant chemistries that have shaped our bodies and our minds.

Perhaps even more importantly, pigs and humans find common cause in the fact that we are both the recently domesticated result of a long tradition of gregarious, playful, tuneful, caring, resourceful, and generally reasonable beings. We are species that can and should get along. Better perhaps than we can, or probably ever will, with any of the great apes who present us with all the problems that beset close relatives everywhere. Pigs and people are just different enough to keep each other guessing, and that looks to me like a very sound basis for any long-term relationship.

The greatest problem at the moment is a communication barrier that exists only because we persist with the stubborn notion that animals lack consciousness. The established position is so entrenched that research that even tries to prove or disprove non-human awareness is discouraged before it can begin. This is astonishingly short-sighted when there is already sufficient evidence out there to make a very good case for animal cognition. All we lack is the courage to allow that some animals really do understand each other instead of just responding in some innate, instinctive fashion.

This acceptance is long overdue where primates are concerned. What I have tried to do here is to suggest that there are good reasons to extend the same presumption to pigs, at least as a working hypothesis. My contention is that present knowledge already shows that pigs can and do distinguish between self and non-self and that they are able to comprehend quite complex circumstances, and to respond to them in meaningful, perhaps even conceptual, ways.

Pigs process thoughts. They understand "if, then" situations, they apply previous experience to novel circumstances, and they interact with their environments, and with each other, as though they are conscious of the consequences.

What more do you want? Pigs with wings?

The paradise of my fancy
is one where pigs have wings

—G. K. CHESTERTON, *Fancies Versus Facts*, 1926

INTRODUCTION: IN A PIG'S EYE

15   The task of organizing the others "fell naturally upon the pigs"; Orwell 1945.

THE FIRST LEG: THE WHOLE HOG

25   Often they band together with other sows; Mohr 1960.

25   Half of this is deep, slow-wave sleep; Kuipers and Whatson 1979.

25   The other half is spent in an unusual, half-awake, drowsy state; Rukebusch 1972.

25   Such ambitious structures are usually for farrowing; Covacevich 1973.

27   It is interesting to know that ancestral pigs were so versatile; Liping, Fortelius, and Pickford 2002.

27   The skulls of entelodonts bear the unmistakable signs of severe head wounds; Lucas and Emry 1996.

28   They have few immediate ancestors in common; Oliver 1993.

29   But its most startling feature in some areas is a smooth coat of beautiful russet color; Grubb 1993.

29   They upset farmers by uprooting banana and papaya trees; Seydack 1991.

30   They are weaned at six months; Breytenbach and Skinner 1982.

30   But in encounters with alien males or outside threats; Jones 1978.

30   All bushpigs also carry the full complement of forty-two pig teeth; Sowls and Phelps 1968.

30   Some of them were even seen hiding underwater; Child 1968.

30   Most sounders proclaim their presence; Phillips 1926.

30   Though no specific studies have been made on their cognition; Maberly 1950.

30   Fossil discoveries suggest that Potamochoerus may be a direct descendant; Harris and White 1979.

30   But in other ways, bushpigs are also astonishingly advanced; Ghiglieri et al. 1982.

31   An omnivorous diet places no particular premium; Herring 1985.

31   On February 25, 1922, Harold J. Cook, a rancher and consulting geologist; Wolf and Mellett 1985.

31   He still worked in the Department of Vertebrate Paleontology; Osborn 1910.

32   He suggested that it might be the third upper molar tooth; Osborn 1922a.

32   On April 25, 1922, in two simultaneous papers; Osborn 1922b, 1922c.

32   He published his opinion on June 24, 1922; Smith 1922.

32   Arthur Smith Woodward at the British Museum of Natural History; Woodward 1922.

32   But on the whole, most people accepted that the finding was; Osborn 1922a.

32   Osborn sent out two new expeditions to Nebraska; Wolf and Mellett 1985.

32   What they found was reported by William King Gregory; Gregory 1927.

33   And the most conspicuous fruit-eaters are the monkeys and apes; Galdikas 1978.

33   Pigs are skilled; Laughlin and Mendl 2000.

33   And this sophisticated spatial memory, which plays a large part in foraging; Laughlin, Huck, and Mendl 1999.

34   It was published in Holland in 1668; Wendt 1959.

34   This "dish" consists of a ring of thickened bone; Cotton 1936.

35   Beneath the eyes are two large, bare "warts"; Ewer 1970.

35   Their young are preyed upon by leopards; D'Huart 1991.

35   If they dig at all, they do so with the lower incisor teeth; Pocock 1943.

36   South of the Sahara, their place in woodland habitats was taken by bushpigs; Cooke 1968.

36   But in 1977 and 1978, three of the most meticulous field workers got together; White and Harris 1977.

36   Veteran paleontologists Tim White, John Harris, and Basil Cooke have identified four major groups; Cooke 1978.

37   It coexisted with, and may even have traveled with, early Homo erectus; White and Harris 1977.

38   These afferent or incoming nerve fibers send signals directly; Pond and Houpt 1978.

38   Nostrils lie in the center of the disc; Bratton, Harmon, and White 1982.

38   The combination of all three makes the pig's snout better equipped; Ewer 1958.

39   Warthogs of the genus *Phacochoerus*—from the Greek *phakos* for "warty" and *khoiros* for "pig"; Cumming 1975.

39   Unlike any other wild pig, warthogs have a tail that comes automatically and vertically to attention; Radke 1991b.

40   The species differ mainly in that incisor teeth are completely lacking; Somers, Penzhorn, and Rasa 1994.

40   Both sexes mate at about eighteen months; Simpson 1964.

40   Then they dive down; Radke 1991a.

40   They are much less dependent than other pigs on water; Clough and Hassam 1970.

42   No relation to the pigs, this bizarre-looking proto-ungulate with ass's ears; Melton 1976.

45   But I noticed that, on such occasions, Hoover seemed uncomfortable; Hacker et al. 1994.

46   As evolving animals develop new ways of exploiting

vegetable foods; Dawkins and Krebs 1979.

46 These give the fungus access to carbohydrates; Watson 1999.

THE SECOND LEG: HIGH ON THE HOG

51 He was 9 feet long and 5 feet tall; Pukite 1999.

53 In tests of acuity, pigs have proven that plastic cards, once nuzzled by them; Hafez and Signoret 1962.

53 As a result, one sniff of the sexual promise carried on a boar's breath; Signoret 1970.

53 But paramount in all such ceremony, in nearly all species, is the saliva; Estes, Cumming, and Hearn 1982.

54 The world in which we live is laced with olfactory information; Ralls 1971.

56 There are *preorbital* glands in front of the eye; Macdonald 1991a.

57 The ideal temperature for most pigs seems to be tropical; Ingram 1965.

57 Unlike us, pigs experience a net water gain under humid conditions; Mount 1968.

57 That said, studies on peccaries show that some of these New World animals; Gabor, Hellgren, and Silvy 1997.

57 *Catagonus wagneri*, the giant or Chacoan peccary; Wetzel, Martin, and Myers 1915.

58 But these factors are less convincing; Eisenberg and McKay 1974.

59 There are reports instead of a preorbital gland; Byers 1978.

59 And it has a very curious origin; Sokolov 1982.

59 Lyle Sowls at the University of Arizona has been studying peccaries; Sowls 1997.

60 Then it has a soothing effect; Byers 1985.

62 They are truly hogs for all seasons; Bigler 1974.

62 This tolerance alone distinguishes them; Green and Grant 1984.

62 The home range varies; Sowls 1974.

62 Submission is signaled by sitting back; Byers and Berkoff 1981.

63 Battles of this kind usually end with the loser kneeling; Schweinsburg and Sowls 1972.

63 Then she picks up the pace; Bissonette 1978.

63 Initially, she protects them; Packard 1981.

63 Under normal circumstances, it is possible to track a herd of collared peccaries; Bissonette 1982.

64 Everything seems to revolve around the shape and nature of the upper canine teeth; Herring 1972.

65 "Why," he asked, "do the opposing teeth need to interlock?" Kiltie 1981b.

65 This, he proposed, is precisely what led to the extraordi-

nary precision; Kiltie 1981b.

65 And, as usual, it was Richard Kiltie who first asked the relevant question; Kiltie 1982.

65 This is enough to provide what ecologists call "limiting similarity"; Emmons 1980.

65 The top third, which required a load of over 1,200 pounds to break; Kiltie 1982.

66 Routine examinations of stomach contents show that both species in the forest; Kiltie 1981c.

66 White-lipped peccaries work in concert; Kiltie 1982.

66 Being part of a large group, however, improves individual safety; Kiltie 1980.

67 Most of the group veered away; Sowls 1997.

67 The big cat felt the force of a dozen interlocking canine-tooth carriers; Roosevelt 1914.

68 Both collared and white-lipped peccaries have three toes on their hind feet; March 1991.

69 Such calls disappear; Sowls 1997.

69 This sound travels far; Hernandez, Barreto, and Ojasti 1995.

70 And in deference to the number of young, these are born earlier; Babbitt and Packard 1990.

70 Once a special teat is selected, the youngster does not drink from any other nipple; Hartsock and Graves 1977.

71 So many domestic piglets end up with lacerations; Fraser 1975.

71 This situation is so extraordinary; Fraser and Thompson 1991.

73 By then, it is too late to compete; DePassille and Rushen 1989.

73 When she is ready to return, she roots all the young into a huddle; Weary et al. 1998.

73 Mothers usually eat their dead infants; Weary et al. 1998.

73 That is an unmistakable call for help; Frädrich 1974.

74 As they grow older, her readiness and their need for protection slowly fade away; Frädrich 1974.

74 Among warthogs, for instance, there may be a certain amount of *allosuckling*; Jensen et al. 1999.

74 In a big herd, does whose young receive donations of milk from other mothers; Ekvall 1998.

74 A study in Uganda showed that juveniles that drink from more than one mother; Jensen 1995.

75 This happens even though they may be only six months old; Schmidt 1991.

75 She leaves her young for longer and longer periods; Bøe 1991.

75 Any attempt to shorten this process in domesticity results in stress; Jensen 1995.

75 Litters still appear at the expected times; Sowls 1984.

77 And very soon we were involved in social play; Smythe 1970.

78 They do all these things as part of coordinated herd activity; Byers 1980.

78 They also have a sound localization threshold; Heffner and Heffner 1989.

78 All noise pushes up their heart rate and blood pressure; Talling et al. 1996.

79 As recently as 1972, it was believed that pigs produced no meaningful sounds; Kiley 1972.

79 A decade later, a study of the voices of suckling pigs produced five different classes; Jensen and Algers 1983.

79 Now, it is grudgingly admitted that pigs are not only extremely vocal animals; Weary and Fraser 1995.

79 They may even have become familiar with her voice while still in the womb; Walser 1986.

79 They communicate their need for both food and heat; Weary, Ross, and Fraser 1997.

79 These are unquestionably reliable signals of real need; Appleby et al. 1999.

79 All such messages become more frequent; Marchant, Whittaker, and Broom 2001.

80 In the young, this manifests itself as "purring"; Sowls 1974.

80 But when these go beyond normal control, they become loud squeals; Schweinsburg and Sowls 1972.

80 Both naturalists divide pig calls into three groups; Frädrich 1965.

80 And at the emotional peak of all possible grunts is the splendid *chant de coeur*; Cumming 1975.

82 They react, it seems, to body size, shape, and facial features; Koba and Tanida 2001.

82 All of this suggests that conclusions such as "visual communication is not considered to be of great importance"; Curtis, Edwards, and Gonyou 2001.

82 To achieve this, the lachrymal bones had to be enormously elongated; Ewer 1958.

82 When grazing with zebras, warthog scanning rates drop dramatically; Scheel 1993.

83 They are also accentuated by an astonishing variety of growths and structures; Frädrich 1965.

84 Some are said even to strengthen this armor; Frädrich 1974.

85 Wild cattle, with majestic sweeps of horn meeting in a heavy boss; Geist 1966.

85 The result was a good show; Barrette 1986.

86 The last of these tall, rangy proto-peccaries became extinct; Wright 1989.

86 The only people who live there are a handful of Indians; Benirschke, Byrd, and Low 1989.

87 Pumas have been seen stalking giant peccaries; Taber 1991b.

87 This is a primitive feature; Sowls 1997.

87 It can truly beat extinction only if the Chaco itself is protected; Taber 1991a.

THE THIRD LEG: HOG HEAVEN

91 This date, however, is linked to the belief that domestication cannot have taken place; Clutton-Brock 1999.

94 Similar evidence has been found at the sites of Jarmo and Argissa-Magula; Clutton-Brock 1979.

94 Crushing dry grain is far simpler than grinding it; Dineley 2002.

95 In 7000 BC, lake dwellings in the Swiss Alps included lean-to sheds; Osborn 1910.

95 [A]nd by 6500 BC, pig remains appear in the domestic garbage; Towne and Wentworth 1950.

95 This discovery is still contentious; Epstein 1969.

95 The bones consist mainly of those from animals less than a year old; Ungnad 1908.

96 Graves of the period often include glazed figurines; Epstein 1971.

96 By the eighteenth dynasty in 1400 BC, even the mayor of El-Kab owned 1,500 pigs; Newberry 1912.

96 When this composite god fell out of favor; Newberry 1912.

96 It seems to have been the fate of all sacred animals; Frazer 1900.

96 It is possible, too, that throughout the warmer areas of Eurasia; Griffin 1992.

97 They have also thrived as feral introductions; Martys 1991a.

97 The lower canines are razor sharp; Groves 1981.

97 In less than a month, the litter are following her around; Martys 1991b.

98 In the tropics, they have even learned how to open coconuts; Spitz and Janeau 1995.

98 They are diurnal when unmolested; Spitz and Janeau 1990.

98 Out of season, squabbles in a sounder are settled with short bouts; Beuerle 1975.

98 As a result wild boars often live to be 20 years old or more; Quenette and Gerard 1992.

99 These have been squeezed back in on themselves; Porter 1993.

101 They are, in the words of veteran hog-watcher William Hedgepeth; Hedgepeth 1988.

103 This means juggling with the genes; Smith, Smith, and Brisbin 1980.

107 He experimented mainly with large white sows; Porter 1993.

108 In his 1524 excursion to Honduras, he led a colorful cavalcade; Towne and Wentworth 1950.

108 But their little herd of Iberian pigs lived happily off the land; Jameson 1907.

108 The rest were left with friendly Indians on the river; Jameson 1907.

110 Any animal brought to market by an Indian was accepted; Young 1844.

110 But the Bowery remained, at least until very recently, as a rundown area; Jameson 1907.

110 Despite this, pigs flourished everywhere; Brisbin 1990.

110 Sometimes it is hard to tell the difference; Graves 1984.

111 John Mayer and Lehr Brisbin of the University of Georgia; Mayer and Brisbin 1991.

111 The implication is that pigs showing signs of domestic changes; Quinn 1968.

112 They were used to wild mountain razorbacks; Laycock 1966.

112 In 1920 George Moore gave away his empty luxury lodge; Wood and Barrett 1979.

112 None other than George Gordon Moore; Singer, Swank, and Clebsch 1984.

113 They lived in the Colorado River delta; Wood and Barrett 1979.

113 Sexual behavior becomes more specific too; Herre and Rohrs 1972.

115 But the most likely derivation appears to have been a misspelling; Hodgson 1847.

115 Others suggested that it might even be an ancestor; Herre and Rohrs 1972.

116 These limited fertile patches are now being rapidly converted; Oliver 1979.

116 Such sounders are not territorial; Oliver 1980.

116 Studies in captivity show that these little pigs sleep very well; Kuipers and Whatson 1979.

116 At one extreme there is the hefty and aggressive wild boar; Oliver 1977.

117 Through all the years in which domestic breeds have been proliferating; Goulding 2001.

117 It is, in short, a conservationist's dream; Yalden 2001.

118 By the reign of Henry VIII, pannage was set at the rate of one penny per pig; Wiseman 2000.

119 Pig-keeping was advised, for instance, for those whose "other occupations furnish a plentiful supply of food"; Bell 1837.

119 Friedrich Engels was startled to find in 1845 that there were flourishing piggeries; Engels 1987.

119 In North Kensington in the mid-nineteenth century, pigs outnumbered people; Richson 1854.

119 In South London, houses were declared "unfit for the keeping of swine"; Malcolmson and Mastoris 2001.

122 In the process they added Ohio, Indiana, Illinois, Michigan, Missouri, Iowa, and Wisconsin; Towne and Wentworth 1950.

123 The pay for those who helped on the drive was about $12 a month; Burnett 1946.

124 But when famous trails like "Old Pike," which ran through the Cumberland Gap; Searight 1894.

125 The British were already improving their native pigs; Spencer 1921.

129 A black sow in the New Forest; Daniel 1803.

129 They could also be easily trained; Hedgepeth 1988.

130 There is hardly any medical discipline that is not already benefiting; Bustad and McClennan 1966.

132 A century ago, G. K. Chesterton exercised himself on the future possibilities of pigs as pets; Chesterton 1908.

THE LAST LEG: HOGGING THE LIMELIGHT

138 He convinces Odysseus of the moral superiority of many animals; Plutarch 1962.

140 What he does do, however, is to "animalize" Lexington; Dahl 1959.

140 On the day of his slaughter, the pig was carried off; Hudson 1919.

140 Like his contemporaries Kafka and Joyce, he used animals to reflect an image; Hudson 1906.

141 They are the Black Pigs of Set; Bleakley 2000.

143 These helpful pigs were presumably incarnations; Sillar, Cameron, and Meyler 1961.

148 She has now worn costumes in the style of all the Hollywood glamour queens; Beard 1981.

149 He decided that there was no better model; Bonera 1990.

149 At Sussex weddings, it was customary to drink to the bride's health; Sillar and Meyler 1961.

150 His direct and conscious heir was George Morland; Sillar and Meyler 1961.

152 Whatever the case, it is hard not to agree with Ryba; Ryba 1983.

152 Joseph Campbell describes the sacrifices; Campbell 1959.

153 So the tracks Persephone follows were those of her mother and herself; Frazer 1900.

156 His stance suggests that of a hunter; Nissensen and Jonas 1992.

157 The pig is a symbol for them both; Frazer 1900.

157 The unlawful, it seems, can be lawful; Mercatante 1982.

160 It is dark colored; Groves 1997.

160 This species has some reddish hair; Macdonald 1990.

160 The Javan warty pig is endangered; Blouch 1988.

160 And from time to time (the last time was in 1987), they aggregate; Caldecott 1991a.

161 No other pig migrates; Caldecott 1991b.

161 It has always been that way; Caldecott and Nyaoi 1985.

161   That means that domestication of wild pigs could have taken place; Groves 1981.

161   There is plenty of evidence of *Sus scrofa;* Oliver 1993.

162   It is a hybrid; Groves 1981.

166   Harris acknowledges the power and persistence of religious traditions; Harris 1978.

167   It defines a man's wealth; Goodale 1995.

167   That is a comfortable ratio; Rappaport 1968.

167   If this doesn't suffice, then the best hope is to throw a truly extravagant pig feast; Vayda, Leeds, and Smith 1961.

167   More sympathetic recent studies have come to appreciate that the importance of pigs; Dwyer 1990.

167   They involve a long series of meetings; Luzbetak 1954.

168   As a rule, sacrifices tend to take place in the early morning; Goodale 1985.

168   Then it can be turned loose; Goodale 1995.

168   In many villages, sunset is announced by high, nasal calls; Dwyer 1990.

170   That means that Melanesians do not get protein as often as pastoralists do; Strathern 1971.

171   This structure is unique; MacKinnon 1981.

171   Local legend insists that babirusas hang by these tusks; Macdonald 1991a.

172   Babirusas occupy a subfamily; Clayton and MacDonald 1999.

172   Males do this vigorously; Leus et al. 1996.

172   Babirusas build nests; Patry, Leus, and Macdonald 1995.

172   In a very small number of aggressive bouts, the combatants stand up; Macdonald et al. 1993.

172   A herd book is being kept by Berlin Zoo; Plasa 1991.

172   The species is rare; Manansang et al. 1996.

172   Babirusa survive, it seems, only as fascinating curiosities; Macdonald 1993.

174   He arranged words; Strutt 1789.

174   On learning that the town's latest pig prodigy was three years old; Boswell 1791.

175   The early romantic poet William Cowper saw the pig performing; Haddon 1996.

175   It was embellished with an elegant frontispiece; Haddon 1996.

175   Bill did all the usual hoops; Hunt 1969.

177   Pigs seem also to have an uncanny sense of timing; Hedgepeth 1988.

177   Cupid was buried in the end in a gold casket; Haddon 1996.

177   The gentleman-farmer Sir Walter Gilbey reported; Gilby 1907.

177   Yet another farmer, the author of a book about pigs, dropped his bucolic guard; Jenkins 1980.

178   "In this respect a republican pig, going where he pleases"; Dickens 1842.

178   William Hedgepeth wonders; Hedgepeth 1988.

178   The story of his bid for freedom led in the end to a fund; Anonymous 2002.

178   Pigs stand out in any farmyard; Koba and Tanida 1999.

179   "In general, there is nothing in the life of a hog"; Youatt and Martin 1835.

180   "It is chiefly because of this unexpectedly favorable relation of subject to method"; Yerkes and Cogurn 1915.

180   Skinner himself said: "It doesn't matter which animal is studied"; Bitterman 1965.

181   Notable among these are an IUCN publication; Klös 1990.

181   One of these brave scholars defines cognitive ethology; Searle 1992.

181   But this interpretation is totally rejected by science; Allen 1998.

183   But there is one interesting line of research that might help us; Croney 2000.

183   This has proved successful in allowing pigs to be tested; Croney et al. 2000.

183   That is true of squirrels and dogs as well; Mendl, Laughlin, and Hitchcock 1997.

183   Before long, the little pig got smart; Held et al. 2000.

183   If the first explanation is true, this means that pigs have what psychologists call "a theory of mind"; Heyes 1993.

184   And, in the end, it is the smarter of the two pigs; Held et al. 2001.

184   The next experiment in Bristol will involve a test; Held et al. 2002.

185   In fact it serves precisely that function; Geist 1966.

185   That's one of the great mysteries about cephalization; Deacon 1997.

186   And they look different; Cerbulis 1994.

186   It is a "chaperon" substance; Holmes 2002.

186   Without it, the hidden mutations were suddenly unmasked; Queltsch, Sangster, and Lindquist 2002.

186   It is now thought that hsp90 may be doing the same thing in plants; Rutherford and Lindquist 1998.

187   Never mind a little gentle learning; McGrew 1998.

Allen, Colin. 1998. Assessing animal cognition: ethological and philosophical perspectives. *Journal of Animal Science* 76: 42–7.

Allen, Mary. 1983. *Animals in American Literature.* Urbana: University of Illinois Press.

Anonymous. 2002. Boar Flees Knife. *The Times,* 16 August.

Appleby, Michael C., Daniel M. Weary, Allison A. Taylor, and Gudrun Illmann. 1999. Vocal communication in pigs: who are nursing piglets screaming at? *Ethology* 105: 881–92.

Babbitt, Kimberley, and Jane M. Packard. 1990. Suckling behaviour of the collared peccary. *Ethology* 86: 102–15.

Baden-Powell, Sir Robert. 1924. *Pig-Sticking or Hog-Hunting.* London: Herbert Jenkins.

Barrette, Cyrille. 1986. Fighting behavior of wild *Sus scrofa. Journal of Mammalogy* 67: 177–9.

Beard, Henry. 1981. *Miss Piggy's Guide to Life.* New York: Knopf.

Bell, Thomas. 1837. *A History of British Quadrupeds.* London.

Benirschke, Kurt, Mary L. Byrd, and Richard J. Low. 1989. The Chaco region of Paraguay. *Interdisciplinary Science Review* 14: 144–7.

Berger, John. 1980. *About Looking.* New York: Pantheon Books.

Beuerle, Wilfried. 1975. Freilanduntersuchungen zum Kampf—und Sexualverhalten des europäischen Wildschwienes. *Zeitschrift für Tierpsychologie* 39: 259–64.

Bierce, Ambrose G. 1906. *The Devil's Dictionary.* New York.

Bigler, William. 1974. Seasonal movements and activity patterns of the collared peccary. *Journal of Mammalogy* 55: 851–5.

Bissonette, John A. 1978. The influence of extremes of temperature on activity patterns of peccaries. *The Southwestern Naturalist* 23: 339–46.

———. 1982. Ecology and social behavior of the collared peccary in Big Bend National Park. *Scientific Monography Series* No. 16. Washington: US National Park Service.

Bitterman, M.E. 1965. Phyletic differences in learning. *American Psychologist* 20: 396–410.

Bleakley, Alan. 2000. *The Animalizing Imagination.* New York: St. Martin's Press.

Blouch, Raleigh A. 1988. Ecology and conservation of the Java warty pig. *Biological Conservation* 43: 295–307.

Bøe, Knut. 1991. The process of weaning in pigs: when the sow decides. *Applied Animal Behaviour Science* 30: 47–59.

Bonera, Franco. 1990. *Pigs: Art, Legend and History.* Boston: Little, Brown.

Bosma, A.A., N.A. de Haan, and A.A. Macdonald. 1991. The current status of cytogenics of the Suidae. *Bongo* 18: 258–72.

Boswell, James. 1791. *The Life of Samuel Johnson.* London.

Bratton, Susan P., Mark E. Harmon, and Peter S. White. 1982. Patterns of European wild boar rooting in the Western Great Smoky Mountains. *Castanea* 47: 230–42.

Breytenbach, G.J., and J.D. Skinner. 1982. Diet, feeding and habitat utilization by bush pigs. *South African Journal of Wildlife Research* 12: 1–7.

Brisbin. I. Lehr, Jr. 1990. A consideration of feral swine as a component of conservation concerns and research priorities for the Suidae. *Bongo* 18: 283–93.

Brooks, Walter. 1968. *The Art of Painting.* New York: Golden Press.

Burnett, Edmund Cody. 1946. Hog raising and hog driving in the region of the French Broad River. *Agricultural History* 20: 91–104.

Bustad, L.K., and R.O. McClennan (eds.). 1966. *Swine in Biomedical Research.* Columbus, Ohio: Battelle Memorial Institute.

Byers, John A. 1978. Probable involvement of the pre-orbital glands in two social behavioral patterns of the collared peccary. *Journal of Mammalogy* 59: 855–56.

———. 1980. Social behavior and its development in the collared peccary. Ph.D. Dissertation at the University of Colorado, Boulder.

———. 1985. Olfaction-related behavior in collared peccaries. *Zeitschrift für Tierpsychologie* 70: 201–10.

Byers, John A., and Marc Berkoff. 1981. Social, spacing and cooperative behavior of the collared peccary. *Journal of Mammalogy* 62: 767–85.

Caldecott, Julian. 1991a. Eruptions and migrations of bearded pig populations. *Bongo* 18: 233–43.

———. 1991b. Monographie des Bartschweines. *Bongo* 18: 54–68.

Caldecott, Julian, and A. Nyaoi. 1985. Sarawak's wildlife. *Sarawak Gazette,* April.

Campbell, Joseph. 1959. *The Masks of God.* New York: Viking Press.

Castro, C.E. 1987. Nutrient effects on DNA and chromatin structure. *Annual Review of Nutrition* 7: 407–21.

Cerbulis, I.G. 1994. Cognitive abilities of the domestic pig. M.S. Thesis, Department of Psychology. Columbus: Ohio State University.

Chatwin, Bruce. 1983. *On the Black Hill.* London: Jonathan Cape.

Chesterton, G.K. 1908. On pigs as pets. In: *The Uses of Diversity,* London: A.P. Watt.

Child, Graham. 1968. Behaviour of large mammals during the formation of Lake Kariba. *Kariba Studies* 1–123.

Child, Graham, Harald H. Roth, and Michael Kerr. 1968. Reproductive and recruitment patterns in warthog populations. *Mammalia* 32: 6–29.

Clayton, Lynn, and David W. MacDonald. 1999. Social organization of the babirusa and their use of salt licks in Sulawesi, Indonesia. *Journal of Mammalogy* 80: 1147–57.

Clough, A., and A.G. Hassam. 1970. A quantitative study of

the daily activity of the warthog in the Queen Elizabeth National Park, Uganda. *East Africa Wildlife Journal* 18: 19–24.

Clutton-Brock, Juliet. 1979. The mammalian remains from the Jericho Tell. *Proceedings of the Prehistoric Society* 45: 135–58.

———. 1999. *A Natural History of Domesticated Mammals*. Cambridge: Cambridge University Press.

Cobbett, William. 1828. *Cottage Economy*. London.

Cooke, H.B.S. 1968. Evolution of mammals in southern continents II: the fossil mammal fauna of Africa. *Quarterly Review of Biology* 43: 236–64.

———. 1978. Suid evolution and correlation of African hominid localities. *Science* 201: 460–63.

Cooke, H.B.S., and A.F. Wilkinson. 1978. Suidae and Tayassuidae. In Vincent J. Maglio and H.B.S. Cooke, *Evolution of African Mammals*. Cambridge: Harvard University Press, 1978.

Cotton, W.B. 1936. Note on the giant forest-hog. *Proceedings of the Zoological Society of London* 687–8.

Covacevich, J. 1973. A nest constructed by wild pigs. *Victorian Naturalist* 93: 25–7.

Croney, Candace Celeste. 2000. Cognitive abilities of domestic pigs. *Dissertation Abstracts International* 62: 1–598.

Croney, Candace Celeste, K.M. Adams, C.G. Washington, and W.R. Stricklin. 2000. Visual, olfactory and spatial clues in the foraging behavior of pigs: indirectly assessing cognitive abilities. *International Society of Applied Ethology Annual Conference in Brazil*.

Cumming, D.H.M. 1975. A field study of the ecology and behaviour of warthog. *Museum Memoir* No. 7, Salisbury, Rhodesia.

Curtis, Stanley E., Sandra A. Edwards, and Harold W. Gonyou. 2001. Ethology and Psychology. In Wilson G. Pond and Harry J. Mersmann (eds.), *Biology of the Domestic Pig*. Ithaca: Cornell University Press, 2001.

Dahl, Roald. 1959. *Kiss, Kiss*. Harmondsworth: Penguin.

Daniel, W.B. 1803. *Rural Sports*. London: Logley and Goatworth.

Darwin, Charles. 1875. *The Varieties of Animals and Plants under Domestication*. London: John Murray.

Dawkins, Richard, and J.R. Krebs. 1979. Arms races between and within species. *Proceedings of the Royal Society of London* 205: 489–511.

Deacon, Terrence W. 1997. What makes the human brain different? *Annual Review of Anthropology* 26: 337–57.

DePassille, Anne Marie B., and Jeffrey Rushen. 1989. Suckling and teat disputes by neonatal piglets. *Applied Animal Behaviour Science* 22: 23–38.

D'Huart, Jean-Pierre. 1991. Monographie des Riesenwaldschweines. *Bongo* 18: 103–18.

Dickens, Charles. 1842. *American Notes*. London.

Dineley, Merryn. 2002. *First Farmers Cultivated a Taste for Horlicks*. The Times, 14 September.

Dubost, G. 1997. Comportements comparés du pécari à lèvres blanches et du pécari à collier. *Mammalia* 61: 313–43.

Dwyer, Peter D. 1990. *The Pigs Ate the Garden*. Ann Arbor: University of Michigan Press.

Dyck, Ian. 1992. *William Cobbett and Rural Popular Culture*. Cambridge: Cambridge University Press.

Eisenberg, J.F., and G.M. McKay. 1974. Comparison of ungulate adaptations in the new world and old world tropical forests. In Valerius Geist and Fritz Walther (eds.), The behaviour of ungulates and its reaction to management. *IUCN Publication* 24 (1974): 144–65.

Ekvall, K. 1998. Effects of social organization, age and aggressive behaviour on allosuckling in wild fallow deer. *Animal Behaviour* 56: 695–703.

Eliot, George. 1857. *Scenes of Clerical Life*. London.

Emmons, L.H. 1980. Ecology and resource partitioning among nine species of African rain forest squirrels. *Ecological Monographs* 50: 31–54.

Engels, Friedrich. 1987. *The Condition of the Working Class in England*. Harmondsworth: Penguin.

Epstein, H. 1969. *Domestic Animals of China*. London: Commonwealth Agricultural Bureau.

———. 1971. *The Origin of the Domestic Animals of Africa*. Edition Leipzig.

Estes, Richard D., David H.M. Cumming, and Gail W. Hearn. 1982. New facial glands in domestic pig and warthog. *Journal of Mammalogy* 63: 618–24.

Ewer, R.F. 1958. Adaptive features in the skulls of African Suidae. *Proceedings of the Zoological Society of London* 131: 135–55.

———. 1970. The head of the forest hog. *East African Wildlife Journal* 8: 43–52.

Frädrich, Hans. 1965. Zur Biologie und Ethologie des Warzenschweines und Berücksichtigung des Verhaltens anderer Suiden. *Zeitschrift für Tierpsychologie* 22: 328–93.

———. 1974. A comparison of behaviour in the Suidae. In Valerius Geist and Fritz Walther (eds.), The behaviour of ungulates and its reaction to management. *IUCN Publication* 24 (1974): 144–65.

———. 1991. Special Volume No. 18 of the journal *Bongo*, Berlin Zoological Garden.

Fraser, David. 1975. The teat order of suckling pigs. *Journal of Agricultural* Science 84: 393–9.

Fraser, David, and B.K. Thompson. 1991. Armed sibling rivalry among suckling piglets. *Behavioural Ecology and Sociobiology* 29: 9–15.

Frazer, James George. 1900. *The Golden Bough*. London: Macmillan.

Gabor, Timothy M., Eric Hellgren, and Nova J. Silvy. 1997. Renal morphology of sympatric Suiformes. *Journal of Mammalogy* 78: 1089–95.

Galdikas, Birute M.F. 1978. Orangutan death and scavenging by pigs. *Science* 200: 68–70.

Garfield, Viola E. (ed.). 1961. Symposium on patterns of land utilization. *Proceedings of the Annual Spring Meeting of the American Ethnological Society.*

Geist, Valerius. 1966. The evolution of horn-like organs. *Behaviour* 27: 175–214.

Geist, Valerius, and Fritz Walther (eds.). 1974. The behaviour of ungulates and its reaction to management. *IUCN Publication* 24: 144–65.

Ghiglieri, M.D., T.M. Butynski, T.T. Struhsaker, and Lysa Leland. 1982. Bush pig polychromatism and ecology in Kibale Forest, Uganda. *African Journal of Ecology* 20: 231–6.

Gilby, Walter. 1907. *Pigs in Health.* Southampton.

Golding, William. 1954. *Lord of the Flies.* London: Faber & Faber.

Goodale, Jane C. 1985. Pigs' teeth and skull cycles. *American Ethnologist* 12: 228–44.

———. 1995. *To Sing with Pigs Is Human.* Seattle: University of Washington Press.

Gould, Stephen Jay. 1991. An essay on a pig roast. In: *Bully for Brontosaurus.* New York: W.W. Norton.

Goulding, M.J. 2001. Possible genetic resources of free-living wild boar. *Mammal Review* 31: 245–8.

Graves, H.B. 1984. Behaviour and ecology of wild and feral swine. *Journal of Animal Science* 58: 482–92.

Green, Galen E., and W.E. Grant. 1984. Variability of observed group sizes within collared peccary herds. *Journal of Wildlife Management* 48: 244–8.

Gregory, William King. 1927. *Hesperopithecus,* apparently not an ape or a man. *Science* 66: 579–81.

Griffin, Donald R. 1976. *The Question of Animal Awareness.* New York: Rockefeller University Press.

———. 1992. *Animal Minds.* Chicago: University of Chicago Press.

Groves, Colin. 1981. Ancestors for the pigs. *Technical Bulletin* No. 3, Research School of Pacific Studies, Australian National University.

———. Taxonomy of wild pigs of the Philippines. *Zoological Journal of the Linnean Society* 120: 163–91.

Groves, Colin P., George Schaller, George Amato, and Kham Khoun Khounboline. 1997. Rediscovery of the wild pig *Sus bucculentus. Nature* 386: 335.

Grubb, Peter. 1993. The Afro-tropical Suids. In William L.R. Oliver (ed.), *Pigs, Peccaries and Hippos.* Gland, Switzerland: IUCN Survival Service Commission, 1993.

Grzimek, B. 1990. *Encyclopedia of Mammals.* New York: McGraw Hill.

Grzimek, B. (ed.). 1972. *Grzimek's Animal Life Encyclopedia.* New York: Van Nostrand Reinhold.

Hacker, R.R., J.R. Ogilvie, W.D. Morrison, and F. Kains. 1994. Factors affecting excretory behaviour of pigs. *Journal of Animal Science* 72: 1455–60.

Haddon, Celia. 1996. *Pigs Are Perfect.* London: Headline.

Hafez, E.S.E. 1962. The adaptation of domestic animals. In E.S.E. Hafez (ed.), *The Behaviour of Domestic Animals.* Baillier, Tindall and Cox, 1962.

Hafez, E.S.E. (ed.). 1962. *The Behaviour of Domestic Animals.* Baillier, Tindall and Cox.

Hafez, E.S.E, and J.P. Signoret. 1962. The behaviour of swine. In E.S.E. Hafez (ed.), *The Behaviour of Domestic Animals.* Baillier, Tindall and Cox, 1962.

Harris, J.M., and T.D. White. 1979. Evolution of Plio-Pleistocene African Suidae. *Transactions of the American Philosophical Society* 69 (2): 1–127.

Harris, Marvin. 1978. *Cannibals and Kings.* London: Collins.

Hartsock, T.G. and H.B. Graves. 1977. Neonatal behaviour and nutrition-related mortality in domestic swine. *Journal of Animal Science* 42: 235–41.

Hedgepeth, William. 1988. *The Hog Book.* Athens, Georgia: University of Georgia Press.

Heffner, Rickye S. and Henry E. Heffner. 1989. Sound localization, use of binaural cues and the superior olivary complex in pigs. *Brain, Behavior and Evolution* 33: 248–58.

Held, Suzanne, Michael Mendl, Claire Devereux, and Richard W. Byrne. 2000. Social tactics of pigs in a competitive foraging task: the informed forager paradigm. *Animal Behaviour* 59: 569–76.

———. 2001. Studies in social cognition: from primates to pigs. *Animal Welfare* 10: 209–17.

Held, Suzanne, Michael Mendl, Keith Laughlin and Richard W. Byrne. 2002. Cognition studies with pigs. *Journal of Animal Science* 80: 10–17.

Hernandez, O.E., G.R. Barreto and J. Ojasti. 1995. Observations of behavioral patterns of white-lipped peccaries in the wild. *Mammalia* 59: 146–8.

Herre, Wolf and Manfred Rohrs. 1972. Animals in captivity. In B. Grzimek (ed.), *Grzimek's Animal Life Encyclopedia.* New York: Van Nostrand Reinhold, 1972.

Herring, Susan W. 1972. The role of canine morphology in the evolutionary divergence of pigs and peccaries. *Journal of Mammalogy* 53: 500–12.

———. 1985. Morphological correlates and masticatory patterns in pigs and peccaries. *Journal of Mammalogy* 66: 603–17.

Heyes, C.M. 1993. Anecdotes, training, trapping and triangulating: do animals attribute mental states? *Animal Behaviour* 46: 177–88.

Hobbes, Thomas. 1651. *Leviathan.* Part One, Chapter 13.

Hodgson, B.H. 1847. On a new form of hog kind or Suidae.

Journal of the Asiatic Society of Bengal 16: 423–8.

Hoff, Benjamin. 1992. *The Te of Piglet*. New York: Dutton.

Holmes, Robert. 2002. Ready, steady, evolve. *New Scientist*, 28 September, 28–31.

Hudson, William Henry. 1906. *Green Mansions*. London.

———. 1919. *The Book of a Naturalist*. London.

Hunt, John. 1969. *A World Full of Animals*. New York: David McKay.

Ingram, D.L. 1965. Evaporative cooling in the pig. *Nature* 207: 415–16.

Jackson, Fatimah Linda Collier. 1991. Secondary compounds in plants and promoters of human biological variability. *Annual Review of Anthropology* 20: 505–46.

Jameson, J. Franklin (ed.). 1907. *Spanish Explorers in the Southern United States*. New York: Scribner.

Jenkins, Adam C. 1980. *A Countryman's Year*. Exeter.

Jensen, P. 1995. The weaning process of free-ranging domestic pigs. *Ethology* 100: 14–25.

Jensen, Per, and B. Algers. 1983. An ethogram of piglet vocalizations during suckling. *Applied Animal Ethology* 11: 227–38.

Jensen, S. Plesner, L. Siefert, J.J.L. Okuri and T.H. Clutton-Brock. 1999. Age-related participation in allosuckling by nursing warthogs. *Journal of Zoology* 248: 443–9.

Jones, M.A. 1978. A scent-marking gland in the bushpig. *Arnoldia* 8: 1–4.

Kelly, Raymond C. 1988. Etoro suidology. In James F. Weiner (ed.), *Mountan Papuans*. Ann Arbor: University of Michigan Press, 1988.

Kiley, M. 1972. The vocalization of ungulates. *Zeitschrift für Tierpsychologie* 31: 171–222.

Kiltie, Richard A. 1980. Application of search theory to the analysis of prey association as an antipredation tactic. *Journal of Theoretical Biology* 87: 201–6.

———. 1981a. Distribution of palm fruits on a rainforest floor: why white-lipped peccaries forage near objects. *Biotropica* 13: 141–5.

———. 1981b. The function of interlocking canines in rain forest peccaries. *Journal of Mammalogy* 62: 459–69.

———. 1981c. Stomach contents of rain forest peccaries. *Biotropica* 13: 234–5.

———. 1982. Bite force as a basis for niche differentiation between rain forest peccaries. *Biotropica* 14: 188–95.

———. 1983. Observations on the behavior of rain forest peccaries in Peru: why do white-lipped peccaries form herds? *Zeitschrift für Tierpsychologie* 62: 241–55.

———. 1989. Peccary jaws and canines. In Kent. H. Redford and John F. Eisenberg (eds.), *Advances in Neotropical Mammalogy*. Gainesville: Sandhill Crane Press, 1989, 151.

Kingdon, Jonathan. 1979. *East African Mammals*. Vol. IIIB.

London: Academic Press.

———. 1997. *Field Guide to African Mammals*. London: Academic Press.

Klös, Heinz Georg (ed.). 1990. Frädrich-Jubiläumsband. *Sitzungsberichte der Tagung über Wildschweine under Pekaris im Zoo Berlin*, July.

Koba, Y., and H. Tanida. 1999. How do miniature pigs discriminate between people? The effect of exchanging cues between a non-handler and their favourite handler on discrimination. *Applied Animal Behaviour Science*. 61: 239–52.

———. 2001. How do miniature pigs discriminate between people? Discrimination between people wearing coveralls of the same colour. *Applied Animal Behaviour Science* 73: 45–58.

Kuipers, M., and T.S. Whatson. 1979. Sleep in piglets. *Applied Animal Ethology* 5: 145–51.

Lamb, Charles. 1824. *Dissertation upon Roast Pig*. London.

Langer, Peter. 1974. Stomach evolution in the Artiodactyla. *Mammalia* 38: 295–314.

Laughlin, Kirsty, and Michael Mendl. 2000. Pigs shift too: foraging strategies and spatial memory in the domestic pig. *Animal Behaviour* 60: 403–10.

Laughlin, Kirsty, Maren Huck, and Michael Mendl. 1999. Disturbance effects of environmental stimuli on pig spatial memory. *Applied Animal Behaviour Science* 64: 169–80.

Laycock, George. 1966. *The Alien Animals*. Garden City, NY: American Museum of Natural History.

Leus, K., K.P. Bland, A.A. Dhondt, and A.A. Macdonald. 1996. Ploughing behavior of babirusa suggests a scent-marking function. *Journal of Zoology* 238: 209–19.

Liping, Liu, Mikael Fortelius and Martin Pickford. 2002. New fossil Suidae from Shanwang, Shandong, China. *Journal of Vertebrate Paleontology* 22: 152–63.

Loizos, Caroline. 1966. Play in mammals. In Caroline Loizos and Peter A. Jewell (eds.), Play, exploration and territory in mammals. *Symposia of the Zoological Society of London*, 1966, No. 18.

Loizos, Caroline, and Peter A. Jewell (eds.). 1966. Play, exploration and territory in mammals. *Symposia of the Zoological Society of London* No. 18.

Lucas, Spencer C., and Robert J. Emry. 1996. Late Eocene Entelodonts from Inner Mongolia, China. *Proceedings of the Biological Society of Washington* 109: 397–405.

Luzbetak, Louis J. 1954. The socioreligious significance of the New Guinea pig festival. *Anthropological Quarterly* 27: 59–80 and 102–28.

Maberly, C.T. Astley. 1950. The African bush-pig: sagacious and intelligent. *Africa Wild Life* 4: 14–18.

Macdonald, Alastair A. 1990. Monographie des Celebes-

Schweines. *Bongo* 18: 39–45.

———. 1991a. Comparative study of functional soft tissue anatomy in pigs and peccaries. *Bongo* 18: 273–82.

———. 1991b. Monographie des Hirschebers. *Bongo* 18: 69–84.

———. 1993. The babirusa: status and action plan summary. In Jansen Manansang, Alastair Macdonald, Dwiatmo Siswomartono, Philip Miller, and Ulysses Seal, *Babirusa: Population and Habitat Viability Assessment.* Gland, Switzerland: IUCN Pigs and Peccaries Specialist Group, 1996.

Macdonald, A.A., D. Bowles, Justine Bell, and Kristin Leus. 1993. Agonistic behaviour in captive babirusa. *Zeitschrift für Saugetierkunde* 58: 18–30.

MacKinnon, John. 1981. The structure and function of the tusks of babirusa. *Mammalian Review* 11: 37–40.

Maglio, Vincent J., and H.B.S. Cooke. 1978. *Evolution of African Mammals.* Cambridge: Harvard University Press.

Malcolmson, Robert, and Stephanos Mastoris. 2001. *The Engish Pig: A History.* London: Hambledon.

Manansang, Jansen, Alastair Macdonald, Dwiatmo Siswomartono, Philip Miller, and Ulysses Seal. 1996. *Babirusa: Population and Habitat Viability Assessment.* Gland, Switzerland: IUCN Pigs and Peccaries Specialist Group.

March, Ignacio J. 1991. Monographie des Weissbartpekaris. *Bongo* 18: 151–20.

Marchant, Jeremy N., Xanthe Whittaker, and Donald M. Broom. 2001. Vocalization of the adult female domestic pig during a standard human approach test and their relationships with behavioural and heart rate measure. *Applied Animal Behaviour Science* 72: 23–39.

Martys, Michael F. 1991a. Monographie der eurasiatischen Wildschweines. *Bongo* 18: 8–20.

———. 1991b. Ontogenie und Funktion der Saugordnung und Rangordnung beim europäischen Wildschwein. *Bongo* 18: 219–32.

Mason, Ian L. 1988. *World Dictionary of Liverstock Breeds.* London: CAB International.

Mayer, John J., and I. Lehr Brisbin, Jr. 1991. *Wild Pigs in the United States.* Athens, Georgia: University of Georgia Press.

McGrew, W.C. 1998. Culture in nonhuman primates? *Annual Review of Anthropology* 27: 301–28.

Meggitt, M.J. 1974. Pigs are our hearts! *Oceania* 44: 165–203.

Melton, Derek A. 1976. The biology of aardvark. *Mammalian Review* 6: 75–88.

Mendl, Michael, Kurt Laughlin and David Hitchcock. 1997. Pigs in space: spatial memory and its susceptibility to interference. *Animal Behaviour* 54; 1491–508.

Mercatante, Anthony S. 1982. *Zoo of the Gods.* New York:

Harper and Row.

Milne, A.A. 1926. *Winnie-the-Pooh.* London: Methuen.

Mohr, E. 1960. *Wilde Schweine.* Neue Brehm-Bucherei No. 247. Wittenburg: Ziemsen Verlag.

Morris, Desmond. 1962. The behaviour of the green acouchi with special reference to scatter hoarding. *Proceedings of the Zoological Society of London* 139: 701–32.

Mount, L.E. 1968. *The Climatic Physiology of the Pig.* Baltimore: Wilkins and Wilkins.

Newberry, P.E. 1912. The pig and cult-animal of Set. *Journal of Egyptian Archaeology* 14: 211.

Nissensen, Marilyn, and Susan Jonas. 1992. *The Ubiquitous Pig.* New York: Abradale Press.

Oliver, William L.R. 1977. The doubtful future of the pigmy hog and the hispid hare. *Journal of the Bombay Natural History Society* 75: 341–55.

———. 1979. Observation on the biology of the pigmy hog. *Journal of the Bombay Natural History Society* 76: 115–42.

———. 1980. *The Pigmy Hog.* Special Scientific Report No. 1, The Jersey Wildlife Preservation Trust.

Oliver, William L.R. (ed.). 1993. *Pigs, Peccaries and Hippos.* Gland, Switzerland: IUCN Survival Service Commission.

Orwell, George. 1945. *Animal Farm.* London: Gollancz.

Osborn, Henry Fairfield. 1910. *The Age of Mammals.* New York: Macmillan.

———. 1922a. *Hesperopithecus,* the anthropoid primate of western Nebraska. *Nature* 110: 281–3.

———. 1922b. *Hesperopithecus,* the anthropoid primate of western Nebraska. *Proceedings of the National Academy of Sciences* 8: 245–6.

———. 1922c. *Hesperopithecus,* the first anthropoid primate found in America. *American Museum Novitates* 37: 1–5.

Packard, J.M., D.M. Dowdell, W.E. Grant, E.C. Hellgren, and R.L. Lochmiller. 1981. Parturition and related behavior of the collared peccary. *Journal of Mammalogy* 68: 679–81.

Packard, J.M., Babbitt, K.J., P.G. Hannon, and W.E. Grant. 1990. Infanticide in captive collared peccaries. *Zoo Biology* 9: 49–53.

Patry, Maurice, Kristin Leus and Alastair A. Macdonald. 1995. Group structure and behaviour of babirusa. *Australian Journal of Zoology* 43: 643–55.

Peck, Robert Newton. 1973. *A Day No Pigs Would Die.* London: Hutchinson.

Peet, Bill. 1965. *Chester, the Worldly Pig.* Boston: Houghton Mifflin.

Phillips, John F.V. 1926. Wild pig at the Knysna: notes by a naturalist. *South African Journal of Science* 23: 655–60.

Plasa, Lutz. 1991. Das Hirscheber-Zuchtbuck. *Bongo* 18: 250–3.

Plutarch. 1962. *Morals: That Brute Beasts Make Use of Reason*. Cambridge: Harvard University Press.

Pocock, R.I. 1943. The external characters of a forest hog and of a babirusa that died in the Society's gardens. *Proceedings of the Zoological Society of London* 113: 36–42.

Pond, Wilson G., and Katherine A. Houpt. (eds.). 1978. *The Biology of the Pig*. Ithaca: Cornell University Press.

Pond, Wilson G., and Harry J. Mersmann (eds.). 2001. *Biology of the Domestic Pig*. Ithaca: Cornell University Press.

Porter, Valerie. 1993. *Pigs: A Handbook to the Breeds of the World*. Sussex: Helm Information.

Potter, Beatrix. 1913. *The Tale of Pigling Bland*. London: Frederick Warne.

———. 1930. *The Tale of Little Pig Robinson*. London: Frederick Warne.

Pukite, John. 1999. *A Field Guide to Pigs*. Helena: Falcon.

Queltsch, Christine, Todd A. Sangster, and Susan Lindquist. 2002. Hsp90 as a capacitor of phenotypic variation. *Nature* 417: 618–24.

Quenette, Pierre-Yves, and Jean-François Gerard. 1992. From individual to collective vigilance in wild boar. *Canadian Journal of Zoology* 70: 1632–5.

Quinn, J.H. 1968. Notes on the exploration of a cave in Arkansas. *Society of Vertebrate Paleontology* 84: 24–6.

Radke, Reihnard. 1991a. Höhlennutzung beim Warzenschwein. *Bongo* 18: 191–218.

———. 1991b. Monographie des Warzenschweines. *Bongo* 18: 119–34.

Ralls, Katherine. 1971. Mammalian scent marking. *Science:* 171: 443–50.

Rappaport, Roy A. 1968. *Pigs for the Ancestors*. New Haven: Yale University Press.

Redford, Kent. H., and John F. Eisenberg (eds.). 1989. *Advances in Neotropical Mammalogy*. Gainesville: Sandhill Crane Press.

Richson, Charles. 1854. *The Observance of the Sanitary Laws Divinely Appointed*. London.

Roosevelt, Theodore. 1914. *Through the Brazilian Wilderness*. New York: Charles Scribner.

Rose, Walter. 1942. *Good Neighbours: Some Recollections of an English Village and Its People*. Cambridge: Cambridge University Press.

Rubel, Paula G., and Abraham Rosman. 1978. *Your Own Pigs You May Not Eat*. Chicago: University of Chicago Press.

Rukebusch, Y. 1972. The relevance of drowsiness in the circadian cycle of farm animals. *Animal Behaviour* 20: 637–43.

Rutherford, Suzanne L., and Susan Lindquist. 1998. Hsp90 as a capacitor for morphological evolution. *Nature* 396: 336–42.

Ryba, Michael. 1983. *The Pig in Art*. London: Orbis.

Scheel, D. 1993. Watching for lions in the grass: the usefulness of scanning and its effects during hunts. *Animal Behaviour* 46: 695–704.

Schmidt, Christian R. 1972. Pigs. In B. Grzimek (ed.), *Grzimek's Animal Life Encyclopedia*. New York: Van Nostrand Reinhold, 1972.

———. 1991. Monographie des Halsbandpekaris. *Bongo* 18: 171–90.

Schweinsburg, Raymond E., and Lyle K. Sowls. 1972. Aggressive behaviour and related phenomena in the collared peccary. *Zeitschrift für Tierpsychologie* 30: 132–45.

Searight, Thomas B. 1894. *The Old Pike*. Uniontown, Pennsylvania.

Searle, J.R. 1992. *The Rediscovery of the Mind*. Cambridge: MIT Press.

Seydack, Armin. 1991. Monographie des Buschschweines. *Bongo* 18: 85–102.

Signoret, J.P. 1970. Reproductive behavior of pigs. *Journal of Reproductive Fertility*, Supplement 11: 105–17.

Sillar, Frederick Cameron and Ruth Mary Meyler. 1961. *The Symbolic Pig*. London: Oliver & Boyd.

Simpson, C. David. 1964. Observations on courtship behaviour in warthog. *Arnoldia* 1: 1–4.

Singer, Francis J., Wayne T. Swank and Edward E.C. Clebsch. 1984. Effects of wild pig rooting in a deciduous forest. *Journal of Wildlife Management* 48: 464–73.

Singer, Peter. 1975. *Animal Liberation*. New York: Random House.

Smith, Grafton Elliot. 1922. *Hesperopithecus:* The Apeman of the Western World. *Illustrated London* News, June 24.

Smith, Michael W., Michael H. Smith, and Lehr Brisbin, Jr. 1980. Genetic variability and domestication in swine. *Journal of Mammalogy* 61: 39–45.

Smythe, N. 1970. On the existence of 'pursuit invitation' signals in mammals. *The American Naturalist* 104: 491–4.

Sokolov, V.E. 1982. *Mammal Skin*. Berkeley: University of California Press.

Somers, Michael J., Barend L. Penzhorn, and Anne E. Rasa. 1994. Home range size, range use and dispersal of warthogs in the Eastern Cape, South Africa. *Journal of African Zoology* 108: 361–73.

Sowls, L.K. 1974. Social behavior of the collared peccary. In Valerius Geist and Fritz Walther (eds.), The behaviour of ungulates and its reaction to management. *IUCN Publication* 24 (1974): 144–65.

———. 1984. *The Peccaries*. Tucson: University of Arizona Press.

———. 1997. *Javelinas and Other Peccaries*. Tucson: University of Arizona Press.

Sowls, L.K., and Robert J. Phelps. 1968. Observations on the African bushpig in Rhodesia. *Zoologica* 53: 75–84.

Spencer, Sanders. 1921. *The Pig*. London: Arthur Pearson.

Spitz, François, and Georges Janeau. 1990. Spatial strategies: an attempt to classify daily movements of wild boar. *Acta Theriologica* 35: 129–49.

———. 1995. Daily selection of habitat in wild boar. *Journal of Zoology* 237: 423–34.

Stapanian, M.A., and C.C. Smith. 1978. A model for seed scatter hoarding: co-evolution of fox squirrels and black walnuts. *Ecology* 59: 884–96.

Steel, David. 1919. *A Lincolnshire Village*. London.

Steig, William. 1968. *Roland the Minstrel Pig*. New York: Harper Collins.

Strathern, Andrew. 1971. Pig complex and cattle complex: some comparisons and counterpoints. *Mankind* 8: 129–36.

Strutt, Joseph. 1789. Pall Mall Pig. In: *The Sports and Pastimes of the People of England*. London.

Taber, Andrew B. 1991a. Monographie des Chaco-Pekaris. *Bongo* 18: 135–50.

———. 1991b. The status and conservation of the Chacoan peccary in Paraguay. *Oryx* 25: 147–55.

Talling, J.C., N.K. Waran, C.M. Wathes, and J.A. Lines. 1996. Behavioural and physiological responses of pigs to sound. *Applied Animal Behaviour Science* 48: 187–201.

Thomas, Dylan. 1954. *Under Milk Wood*. London: Dent.

Towne, Charles Wayland, and Edward Norris Wentworth. 1950. *Pigs: From Cave to Corn Belt*. Norman: University of Oklahoma Press.

Ungnad, A. 1908. Babylonische Miszellen. *Orientalistische Litteratur Zeitung* 2: 533–7.

Vayda, Anthony P., Dorothy Leeds, and David B. Smith. 1961. The place of pigs in Melanesian subsistence. In: V.E. Garfield (ed.), *Proceedings of the Annual Spring Meeting of the American Ethological Society*.

Walser, Elizabeth E. Shillito. 1986. Recognition of the sow's voice by neonatal piglets. *Behaviour* 99: 177–81.

Wardrop, A.E. 1914. *Modern Pig-Sticking*. London: Macmillan.

Watson, Lyall. 1971. *Omnivore*. London: Souvenir.

———. 1976. *Gifts of Unknown Things*. London: Hodder & Stoughton.

———. 1989. *Neophilia*. London: Sceptre.

———. 1997. *Warriors, Warthogs and Wisdom*. London: Kingfisher.

———. 1999. *Jacobson's Organ*. London: Penguin.

Weary, Daniel M., and David Fraser. 1995. Calling by domestic piglets: reliable signals of need? *Animal Behaviour* 50: 1047–55.

Weary, Daniel M., Edmond A. Pajur, Brian K. Thompson, and David Fraser. 1996. Risky behaviour by piglets: a trade-off between feeding and risk of mortality by maternal crushing? *Animal Behaviour* 51: 619–24.

Weary, Daniel M., Peter A. Phillips, Edmund A. Pajur, David Fraser, and Brian K. Thompson. 1998. Crushing of piglets by sows. *Applied Animal Behaviour Science* 61: 103–11.

Weary, Daniel M., Stephen Ross, and David Fraser. 1997. Vocalizations by isolated piglets: a reliable indication of piglet need directed towards a sow. *Applied Animal Behaviour Science* 53: 249–87.

Weiner, James F. (ed.). 1988. *Mountain Papuans*. Ann Arbor: University of Michigan Press.

Wendt, Herbert. 1959. *Out of Noah's Ark*. London: Weidenfeld & Nicolson.

Wetzel, Ralph M., Robert L. Martin and Philip Myers. 1915. *Catagonus*, an 'extinct' peccary, alive in Paraguay. *Science* 189: 379–81.

White, Elwyn Brooks. 1952. *Charlotte's Web*. New York: Harper.

White, Gilbert. 1789. *The Natural History and Antiquities of Selbourne*. London.

White, T.D., and J.M. Harris. 1977. Suid evolution and correlation of African hominid localities. *Science* 198: 13–21.

Wilson, D.S. 1975. On the adequacy of body size as a niche difference. *American Naturalist* 109: 769–84.

Wiseman, Julian. 2000. *The Pig: A British History*. London: Duckworth.

Wodehouse, P.G. 1940. Pig-hoo-o-o-o-ey! In: *Weekend Wodehouse*. 1940. London: Readers Union.

———. 2000. *Pigs Have Wings*. London: Everyman.

Wolf, John, and James S. Mellett. 1985. The role of 'Nebraska Man' in the creation-evolution debate. *Creation-Evolution* 16: 31–43.

Wood, Gene W., and Reginald H. Barrett. 1979. Status of wild pigs in the United States. *Wildlife Society Bulletin* 7: 237–46.

Wood-Gush, D.G.M., and K. Vestergaard. 1991. The seeking of novelty and its relation to play. *Animal Behaviour* 42: 599–606.

Woodward, Arthur Smith. 1922. The Earliest Man? *The Times*, May 22.

Worsley, Peter. 1970. *The Trumpet Shall Sound*. London: Paladin.

Wright, David B. 1989. Phylogenetic relations of Catagonus wagneri. In Kent H. Redford and John F. Eisenberg (eds.), *Advances in Neotropical Mammalogy*. Gainesville: Sandhill Crane Press, 1989.

Yalden, Derek. 2001. The return of the prodigal swine. *Biologist* 48: 259–62.

Yerkes, Robert M., and Charles A. Cogurn. 1915. A study of the behaviour of the pig by a multiple choice method. *Journal of Animal Behaviour* 5: 185–225.

Youatt, William. 1847. *The Pig*. London: Lockwood.

Youatt, William, and W.C.L. Martin. 1835. *The Hog*. New York: Orange Judd.

Young, Alexander. 1844. *Chronicle of the Pilgrim Fathers of the Colony of Plymouth*. Boston: Little & Brown.

5: Bill Ling © Dorling Kindersley. 6: Pot-bellied pig. G K & Vikki Hart © Getty Images. 8–9: Drawings by James Nunn. 10–11: © Dorling Kindersley. 12: © Image Source/CORBIS. 14: Berkshire pig. Gustav Mützel, 1894–1896. Courtesy New York Public Library. 16: James Nunn. 18: Eadweard Muybridge, ca. 1881. Courtesy New York Public Library. 20: Vase from Vatin, Bela Bara, Serbia. Terracotta, Bronze Age. Erich Lessing/Art Resource, NY. 22: Joel Sartore/National Geographic Image Collection. 27: Warthogs and their litter. Beverly Joubert/National Geographic Image Collection. 28: Frank Greenaway © Dorling Kindersley. 35: James Nunn. 37: ©Robert Dowling/CORBIS. 39: James Nunn. 43: © Image 100/Royalty Free/CORBIS. 46: Joel Sartore/National Geographic Image Collection. 47: Roger Phillips © Dorling Kindersley. 48–49: Peccary. Steve Winter/National Geographic Image Collection. 50: S Solum/Photolink © Getty Images. 55: Fototeca Storica Nazionale © Getty Images. 56: Steve Winter/National Geographic Image Collection. 60: James Nunn. 63: James Nunn. 67: Pal Hermansen © Getty Images. 69: James Nunn. 71: © Joe McDonald/CORBIS. 83: © Yann Arthus-Bertrand/CORBIS. 84: Beverly Joubert/National Geographic Image Collection. 87: James Nunn. 88–89: Postcard, private collection. 90: Hermes sacrificing a dog disguised as a pig. Red-figured Attic cup, 510–500 BCE. Kunsthistorisches Museum, Vienna. Erich Lessing/Art Resource, NY. 93: *The Best in the Market*. Henry Atwell Thomas. Lithograph, published by Kimmel & Voigt, 1872. Courtesy Library of Congress. 94: Courtesy New York Public Library. 95: Courtesy New York Public Library. 97: James Nunn. 100: Courtesy New York Public Library. 101: Improved Dorset pigs. E. Hacker and E. Corbet, 1867. Courtesy New York Public Library. 104: *The Market Cart*. Clifford R. James. Mezzotint, 1929. Courtesy Library of Congress. 111: Postcard, private collection. 115: James Nunn. 120: James Nunn. 121: James Nunn. 123: Empire Design Studio. 127: Three cloned miniature pigs from the University of Missouri. © University of Missouri-Columbia/Getty Images. 128: Woman with Vietnamese pot-bellied pig. Birgid Allig © Getty Images. 131: *Ils pullulent les petits cochons, il y en a partout*. Print, 1896. Courtesy New York Public Library. 134–135: *Fiesta Pig*. Andy Warhol. Screenprint printed on Arches 88, 1979 ©2004 Andy Warhol Foundation for the Visual Arts/Artists Rights Society (ARS), New York/Art Resource, NY. 136: One of the Three Little Pigs with Bricklayer. © Blue lantern Studio/CORBIS. 139: Frederic W. Goody. Woodcut, 1904. Courtesy Library of Congress Prints and Photographs Division, Frederic W. Goudy Collection. 142: Varaha. Philip de Bay. © Historical Picture Archive/CORBIS. 145: *Temptation of Saint Anthony*. Hieronymus Bosch, 1490. © The Bridgeman Art Library/Getty Images. 146: © Keline Howard/CORBIS SYGMA. 148: © Julio Donoso/CORBIS SYGMA. 150 (**left**): Ostracon showing a boy and a pig; the pig feeds from a bowl, the boy cries and tries to keep the pig away. Limestone, 1968 BCE, 19th–20th dynasty, from Deir el-Medina. Louvre, Paris. Erich Lessing/Art Resource, NY. 150 (**right**): © C Squared Studios/Getty Images. 151: *The Prodigal Son*. Albrecht Dürer. Engraving, 1496. Courtesy Library of Congress, Bradley Collection. 153: Dummy board. Paint on wood, English, late 18th century. Victoria & Albert Museum, London/Art Resource, NY. 154–155: *Sow*. Alexander Calder. Wire construction, 1928. Museum of Modern Art, NY. © 2004 Estate of Alexander Calder/Artists Rights Society (ARS), New York. Digital Image © The Museum of Modern Art/Licensed by SCALA/Art Resource, NY. 158: Chinese horoscope character. © Dorling Kindersley. 159: James Nunn. 161: Bornean bearded pig. Roy Toft/National Geographic Image Collection. 169: Babirusa (*Babyrousa babyrussa*). Indonesia, Sulawesi. © Art Wolfe/Getty Images. 171: James Nunn. 176: Poster, ca. 1898, by Strobridge Lith. Co., Cincinnatti. Courtesy Library of Congress. 182: Photograph, 1935, by Ringling Bros. and Barnum & Bailey Combined Circus. Courtesy Library of Congress, Prints and Photographs Division. 184: Postcard, private collection. 188–189: Bill Ling © Dorling Kindersley. 190: Pot-bellied pig. © CORBIS. **Endpaper:** Animal Locomotion. Eadweard Muybridge. Photomechanical print (collotype), 1887. Philadelphia: Photogravure Company of New York, plate 673. Courtesy Library of Congress, Prints and Photographs Division.

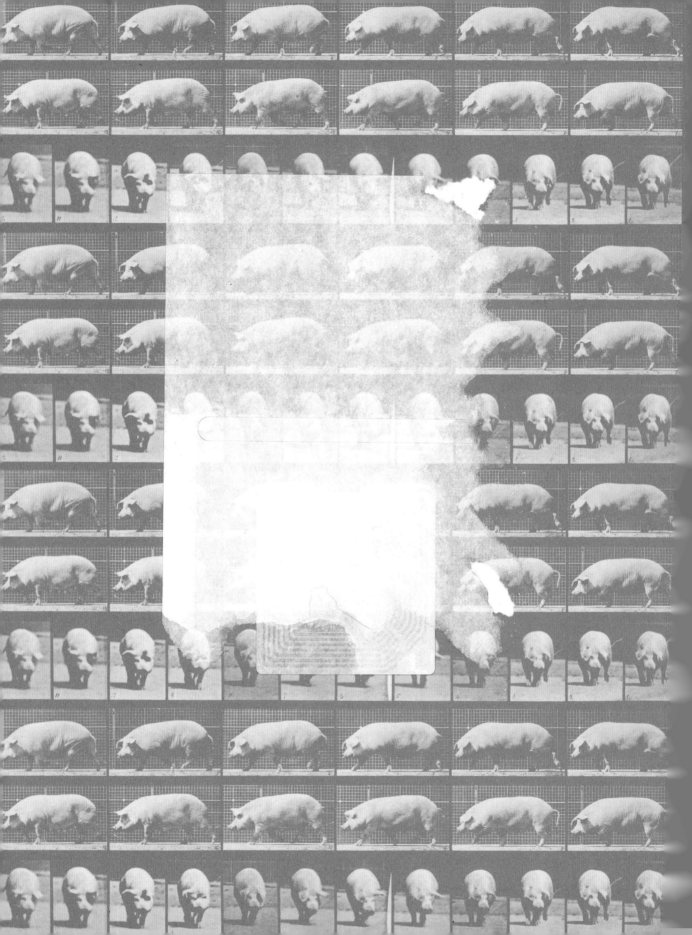